KB090422

Let's
Plunge into Korean Cuisine

You can make Korean food!

궁중음식연구원 지미재

Jimijae, Institute of Korean Royal Cuisine

외국인도 빠져드는
한국밥상

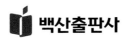 백산출판사

Let's
Plunge into Korean Cuisine

You can make Korean food!

외국인도 빠져드는
한국밥상

추천의 글

축하합니다.

궁중음식연구원에서 수료한 회원들의 모임인 지미재에서 영어로 된 한국음식책을 내게 되었습니다.

지미재는 고 황혜성 선생님께서 궁중의 왕과 대신들의 모임인 기로회처럼 궁중음식을 공부한 제자들을 모아 친목을 도모하는 모임을 만들면서 시작되었습니다.

10명도 안되는 회원들이 이제는 1,000여 명에 달하고 그 활동범위도 넓어져 각지에서 궁중음식을 알리는 주도적인 역할을 하고 계십니다.

지미재 회원들은 음식기능도 뛰어나고 연령대의 폭도 넓어 서로 간에 재능을 나누며 새로운 시대에 맞는 인물로 제 몫을 다하고 있습니다.

회원 중에는 해외에 나가 공부한 유학생들도 많아지니 나름대로 궁중음식연구원이 한식을 세계화하는 데 일조할 수 있겠다고 생각하여 그들을 통하여 영어로 알리는 음식교육을 하게 되었습니다.

음식솜씨가 좋은 회원들과 영어수업이 가능한 젊은 회원들이 서로 힘을 합쳐 돕고 배우며 수업을 진행한 지 4년 반이 지나 이제 공부했던 메뉴를 요리책으로 만들게 되었습니다.

이 책의 출간은 한 사람만의 결과물이 아닙니다.

'음식으로 애국하자'는 캐치프레이즈로 한국음식을 좋아하고 사랑하며, 실제로 만들어서 나누어 먹고, 만드는 법까지 아낌없이 나누는 지미재의 합동작품입니다.

이 책의 저자들은 영어로 가르치는 요리전문가가 아니기에 짜임새가 지금 나오는 전문요리책에 미치지 못할 수도 있습니다.

한국 가정의 음식을 외국인에게 어머니가 쉽게 알려주는 마음으로 만들었기에 누구나 쉽게 만들 수 있을 것입니다. 이 책을 통하여 한국전통음식이 많이 알려져서 한국이 더 잘 알려지는 계기가 되었으면 좋겠습니다.

다시 한 번 포기하지 않고 꾸준히 수업을 진행하며 열과 성을 다해 준 영어요리반 회원들에게 찬사를 보냅니다.

수고하셨습니다.

2015년 3월
궁중음식연구원 이사장 한복려

Congratulations!

This book is created by Jimijae, a group of members who had completed the courses of the Institute of Korean Royal Cuisine.

Jimijae was formed by the late Hye-Seong Hwang, the second holder of the title of Important Intangible Cultural Property. Just like Girohoe, a social meeting group of the king and retired high-rank officials, she wanted to form a group to build fellowship among those who had taken Korean royal cuisine courses.

Starting with only ten members, Jimijae now has over 1,000 members, who are playing a leading role in introducing Korean royal cuisine in diverse fields. Not only Jimijae members have excellent cooking skills but also they have shared their talents despite their age differences, contributing to our society.

As Jimijae attracted more and more members who had studied abroad, it led me to believe that the Institute of Korean Royal Cuisine could contribute to globalizing Korean food. Thanks to Jimijae members' efforts, the institute has offered English cooking classes for foreigners for four and half years.

In class, the older members with excellent cooking skills and the younger members with fluent English collaborated to introduce Korean food to foreigners. This book is based on the teaching materials used in class.

This book is not written by one person. This book is the result of cooperation of Jimijae members who love cooking and sharing Korean food. They have worked hard under the motto: "Contribute to our society by sharing Korean food culture with others."

Since the authors are not professional English cookbook writers, the contents may not be as solid as specialized cookbooks.

However, this book provides easy step-by-step instructions to help foreigners learn and make Korean food easily. I hope that this book will help Korean food and culture be known to more people around the world.

Once again, I would like to praise all members who devoted themselves to the English cooking classes. I highly appreciate their passion and persistence.

Thank you all for your great work.

March 2015

Bokryeo Han, President of the Institute of Korean Royal Cuisine

머리말

　우리는 여러 종류의 채소를 계절에 관계없이 일 년 내내 재래시장이나 대형마트에서 볼 수 있는 시대에
살고 있습니다. 이러한 시대에 태어나서 살기에 우리는 사실 어느 채소가 어느 때 재배되는지, 절기를 알
수 없는 경우가 많습니다. 하지만 선조들의 지혜와 발달된 과학적 지식이 우리에게 알려준 바와 같이 역시
제철에 나온 채소가 영양가도 높고 맛도 뛰어납니다. 이 책을 구성한 집필진도 이에 공감하여 책의 내용을
봄과 여름, 가을과 겨울의 계절별로 나누어보았습니다.

　메뉴 선택의 경우 외국인들이 '한국인의 밥상에는 무엇이 놓여 있을까?' 궁금해 하는 것과 요즘 젊은이
들이 즐겨 먹는 길거리 음식을 위주로 하여 구성하였습니다. 조리법도 이에 어울리게 대대로 내려오는 고
유의 전통방식을 벗어나 앞으로 이 음식과 문화를 이어갈 젊은 세대에게 걸맞게 재해석하여 외국인들이 간
편하게 한국음식을 만들 수 있도록 하였습니다. 기본 맛은 집에서 만든 재래된장, 고추장, 간장이 아닌, 외
국인들이 손쉽게 사서 음식을 만들 수 있도록 시판되는 된장, 고추장, 간장, 쌈장을 사용하였습니다.

　이 책의 내용은 2009년 4월의 봄날 뜻을 같이한 사람들이 시작하여 2014년 6월까지 여러 국가의 외국
인들을 대상으로 수업했던 실기 교재에 이론편을 첨가한 것입니다.

　좋은 책을 만들어야겠다는 무거운 책임감이 저를 짓눌렀지만 진행하는 동안 자기 개발을 위해 헌신하는
젊은 회원들을 보고 제가 도리어 삶의 활력소를 얻었습니다. 궁중음식연구원 지미재의 다재다능한 젊은이
들이 궁중음식, 전통음식에 열정을 가지고 한국음식 지킴이로서 사명을 다하는 모습을 보고 뿌듯했습니다.

궁중음식연구원의 아담하고 견고한 틀 안에 한복려 이사장님의 인재 양성 및 육성이라는 이행목표에 힘입어 이와 같이 궁중음식연구원의 지미재 회원들이 연구원에서 외국인을 대상으로 수업할 수 있었던 것은 집필진의 행복이었습니다. 또한 누구보다도 강한 열정으로 우리의 음식을 배우던 외국인 학생들에게 격려의 장학금도 전달해 주신 한복려 이사장님께 진심으로 감사드립니다.

이 책을 냄으로써 궁중음식연구원 지미재는 한 단계 도약하는 계기가 되었으며 회원들이 만들어온 이러한 토대는 앞으로 한국음식의 새로운 역사를 만들게 될 것이라 믿어 의심치 않습니다.

한국문화를 음식을 통해 배우려는 외국인, 음식과 영어에 취미가 있는 한국인 학생, 외국인을 대상으로 한국음식을 가르치시는 선생님들께 조금이나마 보탬이 되기를 바랍니다.

2015년 3월

제13대 지미재 회장 김매순

Preface

Today we can find all kinds of vegetables in the market all year around, so we are often un-aware of vegetables in season. However, vegetables in season are most nutritious and tasty, as indicated by both our ancestors' wisdom and modern science. We, the authors, totally agree with it and that's why the contents of this book is divided by four seasons.

The food items in this book are also two-fold: Korean meal items interesting to foreigners and street food popular with young people. We have done our best to provide not only reader-friendly but also practical recipes. We have simplified the traditional recipes as much as possible so that foreigners and young people can cook Korean food themselves easily. For flavor, this book uses ingredients and seasonings that readers can buy readily at stores.

From April 2009 to June 2014, the Institute of Korean Royal Cuisine offered the English class-es teaching how to cook Korean food to foreigners. This book is based on our textbook for the classes, and includes more theoretic explanations.

At the beginning, I was under heavy pressure to publish a book successfully. However, soon after I was energized by talented young Jimijae members who committed to developing this book. They were striving to promote Korean food with great passion. I was so proud of them and grateful.

We highly appreciate solid supports from the Institute of Korean Royal Cuisine. With our president Bok-Ryeo Han's leadership, passion for cultivating young talents, and commitment to education, we could initiate the English cooking class program for foreigners and continue with the program for four and half years. I am enormously grateful to president Han who even gave scholarships to foreign students learning Korean food and culture with their passion. I believe that this book was an opportunity for Jimijae to make a big step towards a new chapter in the history of Korean food.

Hopefully, this book will be useful to all foreigners exploring Korean culture through food. Also, I believe that this book will be helpful to Koreans interested in cooking and English, including those who teach Korean cooking to foreigners.

March 2015

Maesoon Kim, The 13th President of Jimijae

CONTENTS

외국인도 빠져드는 한국밥상

15

가을 / 겨울

별식

외국인도 빠져드는 한국밥상

우리 음식의 기본준비
Basics of Korean Food

● 양념

'양념'이란 '먹어서 몸에 약처럼 이롭기를 바라는 마음으로 여러 가지를 고루 넣어 만든다'는 뜻이 담겨 있다. 음식을 만들 때는 재료가 지닌 고유의 맛을 살리면서 음식의 특유한 맛을 내기 위해 여러 가지 재료가 사용되며 이러한 재료들을 양념이라 한다. 한국음식은 한 가지 음식을 만드는 데 대여섯 가지의 양념이 들어가 독특한 맛을 낸다.

양념은 조미료와 향신료로 나눌 수 있다. 조미료는 짠맛, 단맛, 신맛, 매운맛, 쓴맛의 기본적인 맛을 내며 한국음식은 조미료들을 적당히 혼합하여 알맞은 맛을 낸다. 소금, 간장, 고추장, 된장, 식초, 설탕 등이 조미료이다. 향신료는 좋은 향을 지녔거나 매운맛, 쓴맛, 고소한 맛 등을 내며, 식품 자체가 지닌 좋지 않은 향을 없애거나 감소시키며, 특유의 향으로 음식의 맛을 더욱 좋게 한다. 향신료에는 생강, 겨자, 후추, 고추, 참기름, 들기름, 깨소금, 파, 마늘 등이 있다. 한국음식은 다른 나라 음식보다 참기름, 깨소금, 파, 마늘, 고춧가루의 사용량이 많아 독특한 맛의 차이가 난다.

음식의 가장 기본적인 맛은 '짜다' 또는 '싱겁다'의 간을 내주는 '짠맛'을 내는 조미료이다. 소금과 장류인 간장, 된장, 고추장, 젓갈류인 새우젓 등이 쓰인다. 음식에 따라 가장 맛있게 느끼는 간은 농도가 다르다. 맑은국은 1% 정도, 맛이 진한 토장국이나 건지가 많은 찌개는 간의 농도가 더 높아야 하고, 찜이나 조림 등 고형물의 간은 더욱 강해야 맛있게 느껴진다.

● Yangnyeom

The word 'yangnyeom' has the meaning of 'making food with a variety of ingredients, hoping that it will be good for health as medicine.' When cooking, various ingredients can be used to create the unique flavor of a food as well as to enrich the original flavor of the main ingredients. Such ingredients are called 'yangnyeom.' The unique flavor of a Korean food is often created through the combination of several yangnyeoms.

Yangnyeoms fall into two categories: seasoning and flavor enhancer. Generally, seasonings are used to create basic flavors such as salty, sweet, sour, spicy, and bitter tastes. A right mixture of seasonings is essential to make a good flavor of a Korean food. Korean seasonings include: salt, soy sauce, red chili paste, soybean paste, vinigar, and sugar. Flavor enhancers either have a pleasant scent or create a spicy, bitter or savorly taste. They reduce or eliminate the unpleasant scent of ingredients and make food taste better with their unique fragrance. Korean flavor enhancers include: ginger, mustard, black pepper, red chili, sesame oil, perilla oil, ground sesame seeds, green onion, and garlic. Especially, Korean food uses sesame oil, ground sesame seeds, green onion, garlic, and red chili powder very often compared to other countries' food.

The most basic seasonings are those that adds a salty taste. Typical Korean salty seasonings are salt, soy sauce, soybean paste, red chili paste, and salted shrimp. The proper level of seasoning depends on the food. For a clear soup, the concentration is around 1%. For a thick soup or stew with lots of ingredients, the concentration should be higher. For steamed or braised food, the solid ingredients should be seasoned more strongly to make them taste better.

• 소금

음식의 가장 기본적인 맛을 내는 조미료이다. 종류로는 호렴, 재염, 재제염, 식탁염, 맛소금 등이 있다. 호렴은 입자가 굵으며 장을 담그거나 채소나 생선을 절일 때 사용된다. 호렴에서 불순물을 제거한 재염은 간장이나 채소나 생선의 절임용으로 쓰인다. 재제염은 희고 입자가 고운 소금으로 음식에 직접 간을 맞출 때 쓰이며 가정에서 가장 많이 사용한다. 식탁염은 이온교환법에 의해 만들어진 고운 입자의 소금이고, 맛소금은 정제염에 글루타민산을 첨가한 것이다.

• Salt

Salt creates the most basic flavor of foods. There are several types of salts: horyeom, jaeyeom, jaejeyeom, table salt, and seasoned salt. Horyeom has large grains and is used for making soy sauce or for salting vegetables or fish. Jaeyeom is made of horyeom by eliminating impurities. Jaejeyeom has white fine grains and is used for spicing up food, mostly at home. Table salt has fine grains and made chemically through the ion exchange method. Seasoned salt is made by adding glutamic acid to refined salt.

• 간장

콩으로 만든 발효식품으로 음식의 맛을 내는 중요한 조미료이다. 요즘은 집에서 장을 담그지 않고 공장에서 제조하여 시판되는 제품을 쓰는 가정이 많아졌다. 음식에 따라 간장의 종류를 구별하여 써야 한다. 국 · 찌개 · 나물 등에는 국간장을 쓰고, 조림 · 포 · 볶음 등의 조리와 육류의 양념에는 진간장을 쓴다. 간장은 조미료만이 아니라 상에 올리는 초간장, 양념간장에도 사용하며 초간장에는 간장에 식초를 넣고 양념간장에는 고춧가루, 다진 파 · 마늘을 넣는다.

• Soy Sauce

Soy sauce is made by fermenting beans and it is a very useful seasoning in Korean food. These days most people buy it at a store, instead of making it at home. It is important to use the right type of soy sauce depending on the dish. For example, 'gukganjang' (clear soy sauce) is appropriate for soup, stew, and seasoned vegetable, while 'jinganjang' (dark soy sauce)

 matches braised food, dried food, stir-fried food, and seasoned meat. Mixed with other ingredients, soy sauce can be served as dipping sauce; 'choganjang' is soy sauce mixed with vinegar and 'yangnyeomganjang' is soy sauce mixed with red chili powder, chopped green onion and garlic.

• 된장

된장은 조미료뿐만 아니라 단백질의 급원 식품이기도 하다. 재래식으로는 간장을 떠내고 남은 건더기가 된장이다. 근래에는 공업적으로 된장을 만드는데 콩과 밀을 섞어 발효시켜서 만든다. 된장은 주로 토장국이나 된장찌개의 맛을 내는 데 사용하고, 상추쌈이나 호박잎쌈에 곁들이는 쌈장의 재료가 된다.

• Soybean Paste

Soybean paste is a good source of protein as well as a seasoning. Traditionally, soybean paste was made at home; the remaining solid ingredients after making soy sauce are soybean paste. These days it is manufactured at a factory by mixing and fermenting beans and wheat. Soybean paste is mostly used for 'tojang-guk' and 'doenjang-jjigae'. It is also often used for 'ssamjang' (a mix of soybean paste and other seasoning ingredients), which is usually served together with lettuce-wrapped rice or green squash leaf-wrapped rice.

• 고추장

고추장은 한국 고유의 식품으로 간장, 된장과 함께 세계에서 유일한 맛을 내는 조미료이다. 고추장은 고춧가루, 메줏가루, 곡물가루, 소금, 물을 넣고 발효시켜 만든다. 탄수화물이 가수분해되어 생긴 단맛과 콩단백에서 오는 아미노산의 감칠맛, 고추의 매운맛, 소금의 짠맛이 조화를 이룬 조미료이며 기호식품이다. 고추장은 토장국이나 고추장찌개에 맛을 내고, 생채나 숙채, 조림, 구이 등에 사용된다. 또한 초고추장이나 비빔밥 또는 비빔국수에 넣는 볶음고추장도 만든다.

• Red Chili Paste

Red chili paste is a very unique seasoning of Korean food. Along with soy sauce and soybean paste, it creates Korean food's distinct flavors. Red chili paste is made by fermenting a mix of red chili powder, fermented soybean powder, grain powder, salt, and water. It provides a harmony of various tastes: a sweet flavor of carbohydrates decomposed in water, a savory taste of amino acids from soy protein, a hot taste of red chili, and a salty taste of salt. Red chili paste is used for various Korean dishes: 'tojang-guk', 'gochu-jang-jjigae' (red chili paste stew), 'saengchae' (seasoned vegetables), 'jorim' (braised food), and 'gui' (roasted food). Red chili paste with vinegar, which is called 'chogochujang,' is used as a dipping sauce, and stir-fried red chili paste is used for 'Bibimbap' (mix with rice, vegetables, and meat) and 'Bibim-guksu' (spicy noodles).

• 새우젓

새우젓은 작은 새우를 소금에 절인 젓갈로서 김치에 가장 많이 쓰인다. 소금 간보다 감칠맛을 내므로 국, 찌개, 나물 등의 간을 맞출 때도 쓰인다. 특히 호박, 두부, 돼지고기로 만든 음식과 맛이 잘 어울린다.

• Salted Shrimp

Salted shrimp is widely used for Kimchi. Since it is not only salty but also savory, it is also often used for soup, stew and seasoned vegetables instead of salt. In particular, it goes well with dishes whose ingredient is green squash, Dubu or pork.

• 설탕 · 꿀 · 조청

설탕은 사탕수수나 사탕무의 즙을 농축시켜 만드는데 순도가 높을수록 단맛이 강하고 산뜻하다. 당밀분을 포함한 흑설탕보다 정제도가 높은 흰설탕이 단맛이 가볍다.

조청은 곡류를 엿기름으로 당화시켜 오래 고아서 걸쭉하게 만든 묽은 엿으로 누런색이고 독특한 향이 있다. 한과나 밑반찬용 조림에 많이 쓰인다.

꿀은 꽃의 꿀과 꽃가루를 모아서 만들며 가장 오래 이용한 천연감미료이다. 꿀은 꿀벌의 종류와 밀원이 되는 꽃의 종류에 따라 색과 향이 다르다. 음식의 감미료보다는 과자 · 떡 · 정과 등에 쓰인다.

• Sugar, Honey, and Grain Syrup

Sugar is made of the concentrated juice of sugar canes or sugar beets. The higher the purity is, the stronger the sweet taste and the refreshing feel are.

Compared to the black sugar with molasses, the white sugar with higher purity tastes more light. Grain syrup is a thick yellow taffy with a scent. It is made of grains sweetened with malt; it should be boiled for a long time. Featuring a unique flavor, grain syrup is often used for Korean tranditional cookies and side dishes which are boiled in soy sauce. Honey is the oldest natural sweetener. The color and taste of honey depends on the type of bees and flowers. Honey is mostly used for Korean traditional cookies, rice cakes, and fruit preserved in honey, rather than a meal.

• 식초

식초는 음식의 신맛을 내는 조미료이다. 신맛은 음식에 청량감과 식욕을 증진시키고 소화액의 분비를 촉진하여 소화 흡수를 돕는다. 한국음식은 대개 차가운 음식에 식초를 넣는다. 생채와 겨자채, 냉국 등에 넣어 신맛을 낸다. 식초는 녹색의 엽록소를 누렇게 변색시키므로 푸른색 나물이나 채소에는 먹기 직전에 넣어 상에 낸다.

• Vinegar

Vinegar creates a sour taste. It is known that a sour taste whets refreshing feel and appetite and helps digestion through increasing the secretion of digestive juices. In Korean food, vinegar is usually used for cold dishes: 'saengchae' (seasoned vegetables), 'gyeoja-chae' (vegetables with mustard dressing), 'Naeng-guk' (clear cold soup). Since vinegar can make chlorophyll of vegetables turn yellow, for green vegetables, vinegar should be used right before they are served.

• 파

파는 자극성 냄새와 독특한 맛으로 향신료 중에 가장 많이 쓰인다. 굵은 파, 실파, 쪽파, 세파 등의 종류가 있고 나오는 계절이 다르다. 파의 흰 부분은 다지거나 채 썰어 양념으로 쓰이며, 파란 부분은 채 썰거나 크게 썰어 찌개나 국에 넣는다. 파의 매운맛을 내는 성분은 가열하면 맛이 부드러워지고 단맛이 강해진다.

Green Onion

Featuring a strong scent and unique flavor, green onion is used most often among all the flavor enhancers of Korean food. Green onion has four types with different seasons: green onion, small green onion, scallion, and thin green onion. The white part of green onion is usually chopped finely

or julienned to be used as a spice, while the green part is either julienned or cut in chunks for a stew or a soup. Heating makes the hot flavor of green onion softened and sweetened.

마늘

마늘은 독특한 자극성 맛과 향기가 있으며, 특히 육류 요리에 많이 쓰인다. 나물이나 김치 또는 양념장에 곱게 다져서 쓰고, 동치미나 나박김치에는 채 썰거나 납작하게 썰어 넣는다.

Garlic

Garlic has pungent taste and scent. It is often used for meat dishes. Also, minced garlic is used for kimchi, 'namul' (Korean seasoned vegetables), and 'yangnyeomjang' (a mix of soy sauce paste and other seasoning ingredients). For 'dongchimi' (waterly radish kimchi) and 'Nabak-kimchi' (waterly kimchi), julienned or sliced garlic is used.

생강

생강은 매운맛과 쓴맛을 내며 어패류나 육류의 비린내를 없애는 작용을 한다. 생선이나 육류에는 처음부터 넣는 것보다 재료가 어느 정도 익은 후에 넣는 것이 효과적이다. 생강은 음식에 따라 갈아서 즙만 넣고 곱게 다지거나 채로 썰거나 얇게 저며 사용한다. 향신료뿐 아니라 음료나 한과를 만들 때도 많이 쓰인다.

Ginger

Ginger has both hot and bitter tastes. It eliminates the unpleasant smell of fish and meat and it is more effective to add ginger when fish or meat is cooked slighly. Depending on the dish, ginger is ground, minced, julienned, or sliced. It is also used for Korean traditional cookies and drinks.

• 후추

매운맛을 내는 향신료로서 생선이나 육류의 비린내를 제거하고 음식의 맛과 향을 좋게 하고 식욕도 증진시킨다.

• Black Pepper

Black pepper has spicy flavor and eliminates the unpleasant smell of fish and meat. It also makes flavor and scent better and whets appetite.

• 고추

한국음식의 매운맛을 내는 데는 주로 고추가 쓰인다. 고추는 품종이나 산지, 건조법에 따라 맛의 차이가 있으며, 용도에 따라 굵은 고춧가루, 중간 고춧가루, 고운 고춧가루로 나누어 빻는다. 실고추는 나박김치에, 고춧가루는 김치나 깍두기에, 고운 고춧가루는 일반 조미용과 고추장에 적당하다.

• Red Chili

Korean food's hot taste is primarily created by red chili. Red chili's taste is different depending on the variety, the area of production, and the drying method. There are three types of ground red chili: coarse red chili powder, medium red chili powder, and fine red chili powder. Medium red chili powder is used for kimchi and 'kkakdugi' (diced radish kimchi); fine red chili powder is ideal for general seasoning and red chili paste; and red chili strip is appropriate for 'Nabak-kimchi' (waterly kimchi).

• 겨자

갓의 씨를 가루로 빻은 것으로 따뜻한 물로 개어 공기 중에 방치하면 매운맛이 난다. 시중에서 파는 연겨자는 발효해서 튜브에 넣어 바로 쓰기에 편하다.

• Mustard

Mustard is made by grinding the seeds of leaf mustard, mixing the powder with warm water, and leaving outside until it tastes spicy. The soft mustard sold in a tube is fermented mustard and is convenient to use immediately.

• 참기름

한국음식에 가장 많이 쓰이는 기름으로 참깨를 볶아서 짠다. 참기름은 튀김기름 으로는 쓰이지 않으며, 나물 무칠 때와 고기 양념, 약과 · 약식 등 향을 내기 위한 거의 모든 음식에 넣는다.

• Sesami Oil

Sesame oil is the most used oil for Korean food. It is made by roasting sesame seeds and pressing them. Sesame oil is not used for any frying foods, but it is used for most Korean foods that need a savory scent: 'Namul' (Korean seasoned vegetables), marinade for meat, Korean traditional cookies, and so on.

• 들기름

들깨를 볶아서 짠 것으로 참기름과는 다른 고소하고 독특한 냄새가 난다. 김에 발라 굽거나 나물을 무칠 때, 그대로 갈아서 즙을 만들어 나물을 무치거나 냉국과 된 장국에 넣기도 한다.

• Perilla Oil

Perilla oil is made by roasting perilla seeds and pressing them. It has unique savory flavor and scent different from sesame oil. Perilla oil is used for toasting dried laver, seasoning vegetables, and spicing up 'Naeng-guk' (clear cold soup) and 'Doenjang-guk' (soybean paste soup).

---

* 위의 상품은 수업에 사용한 제품입니다.

* The products above were used during class.

● 고명

'고명'이란 음식을 보고 아름답게 느껴서 먹고 싶은 마음이 들도록 모양과 색을 좋게 하기 위해 장식하는 것을 말한다. 한국음식의 색깔은 오행설(五行設)에 바탕을 두어 붉은색, 녹색, 노란색, 흰색, 검은색의 오색이 기본이다. 붉은색은 붉은 고추 · 실고추 · 대추 · 당근 등으로, 녹색은 미나리 · 실파 · 호박 · 오이 등으로, 노란색과 흰색은 달걀의 황 · 백지단으로, 검은색은 석이버섯 · 목이버섯 · 표고버섯 등을 사용한다. 이외에 잣, 은행, 호두 등 견과류와 고기완자 등도 고명으로 쓰인다.

● Gomyeong

'Gomyeong' means decorating food to enhance the color and shape so that people want to eat it, attracted by its beauty. Korean food has the five basic colors based on the Yin−Yang and Five Element theory: red, green, yellow, white, and black. Red color is created by using red chili, red chili powder, red chili strips, jujube, and carrot. Green color is created by using Korean watercress, small green onion, green squash, and cucumber. Yellow and white colors are created by egg garnish. Black color is created by using black mushrooms. Additionally, nuts such as pine nuts, gingko nuts, and walnuts, and meat balls can be used as garnish.

● 달걀지단

달걀을 흰자와 노른자로 나누어 소금을 각각 넣고 기름 두른 팬에 불을 약하게 해서 달걀을 붓고 얇게 펴서 부친다. 각각 용도에 맞게 썰어 쓰는데 채 썬 지단은 나물이나 잡채, 골패형(직사각형)과 완자형(마름모꼴)은 국이나 찜 · 전골 등에 쓰인다. 줄알이란 뜨거운 장국(국물)이 끓을 때 풀어 놓은 달걀을 줄 긋듯이 넣어서 부드럽게 엉기게 하는 것을 말한다.

● Jidan (Egg Garnish)

'Jidan' is made by separating the egg yolk from the egg white, spreading each of them on a greased pan on low heat after adding salt, and frying it. When done, it should be cut into the right shape depending on the food. Julienned one is used for 'Namul' (Korean seasoned vegetables) or 'Japchae' (starch noodles with various vegetables and meat); rectangular or diamond-shape one is used for soup, braised food, or stew. When a hot clear soup or broth starts to boil, putting stirred eggs into the soup as drawing lines creates a certain pattern of tangled eggs. It is called 'Jul-al.'

• 고기완자

쇠고기의 살을 곱게 다져서 소금으로 양념하여 고루 섞어 둥글게 빚는다. 빚은 완자에 밀가루를 입히고 풀어 놓은 달걀물을 입혀 팬에 기름을 두르고 굴리면서 전체를 익힌다. 면, 전골, 신선로의 고명으로 쓰이고 완자탕의 건더기로도 쓰인다.

• Meatball

Meatball is made by mincing a piece of beef, seasoning it with salt, shaping it into a ball with hands, coating the balls with flour and stirred eggs, and rolling the balls on a greased pan until they are well-cooked. Meatball is used as garnish for noodles, stew, and 'Sinseonro' (hot pot of Korean royal cuisine). It is also used for wonton soup.

• 고기 고명

쇠고기를 곱게 다져서 양념하여 볶아 식힌 후 곱게 다져서 국수장국이나 비빔국수의 고명으로 쓴다. 쇠고기를 가늘게 채 썰어 양념하여 볶은 후 떡국이나 국수의 고명으로 얹기도 한다.

• Meat Garnish

Meat garnish is made by mincing a piece of beef, seasoning it with salt, stir-frying it on a greased pan, and letting it cool down. Chopped finely,

it is used as garnish for clear noodle soup and spicy noodles. For 'Tteok-guk' (rice cake soup) and Korean traditional noodles, julienned beef that is seasoned and stir-fried can be used as garnish.

• 표고버섯

마른 표고버섯을 한 번 헹군 후, 미지근한 물이나 찬물을 잠기게 부어 기둥 부분까지 부드러워지면 기둥을 떼고 용도에 맞게 썬다. 전을 부칠 때는 작은 크기로 쓰고, 채로 써는 것은 두꺼우면 얇게 저며 썬 다음 채로 썬다. 고기 양념장으로 양념하여 볶아서 쓴다.

• Pyogo Mushroom

Pyogo mushrooms are prepared by rinsing dried pyogo mushrooms in running water, letting them soak in warm or cold water until the caps get softened, then taking the caps off, and cutting the mushrooms properly depending on the food. For 'jeon' (Korean cooking method of panfrying fritters),

pyogo mushrooms are cut into small pieces. When pyogo mushrooms need to be julienned, the easier way is slicing them first. Pyogo mushrooms can be stir-fried with meat marinade.

• 석이버섯

석이버섯은 뜨거운 물에 불려 양손으로 비벼서 안쪽의 이끼를 말끔히 벗겨낸 후, 여러 번 물에 헹구어 낸다. 큰 것은 바깥에 안쪽에는 작은 것을 놓고 말아서 채 썰거나 다져서 쓴다. 보쌈김치, 국수, 잡채, 떡 등의 고명으로 쓴다.

• Stone Ear Mushroom

Stone ear mushrooms are prepared by letting them soak in hot water, removing the moss inside by rubbing by hands, and then rinsing them several times. Stone ear mushrooms are generally julienned. They are used as garnish for 'bossam kimchi' (kimchi for boiled pork belly), 'japchae' (starch noodles with various vegetables and meat), noodles, rice cakes, and other dishes.

• 실고추

말린 고추를 갈라 씨를 털어내고 꼭꼭 말아서 곱게 채 썬다. 나물이나 국수의 고명, 김치에 많이 쓰인다.

• Silgochu (Thinly Cut Dried Red Chilies)

Silgochu are made by removing seeds from dried red chili rolling the red chili' and making them finely julienned. Silgochu are used as garnish for 'Namul' (Korean seasoned vegetables), noodles, and kimchi.

• 붉은 고추 · 풋고추

말리지 않은 붉은 고추와 풋고추를 반으로 갈라 씨를 빼서 채로 썰거나 완자형, 골패형으로 썰어 쓴다. 잡채나 국수의 고명으로 쓰인다.

• Red Chili and Green Chili

Red chilies and green chilies are used as garnish for noodles and 'Japchae' (starch noodles with various vegetables and meat). As garnish, red chilies and green chilis are prepared by cutting fresh chilies in half, removing seeds, and then cutting the chilies into rectangular shape or making the chilies julienned.

• 실파 · 미나리

가는 실파나 미나리 줄기를 데쳐서 용도에 맞게 썰어 찜, 전골이나 국수의 고명으로 쓴다. 끓는 물에 소금을 넣고 데쳐 찬물에 헹궈 식혀야 푸른색을 살릴 수 있다.

• Small Green Onion and Korean Watercress

Small green onion and Korean watercress are used as garnish for steamed foods, stews, and noodles. As garnish, they are prepared by blanching them in boiling water with salt and rinsing them in cold water. Rinsing in cold water helps their green color preserved.

• 잣

뽀족한 쪽의 고깔을 떼고 통째로 쓰거나 길이로 반 갈라서 비늘잣으로 쓰거나 다져서 잣가루로 쓴다. 잣가루를 만들 때는 도마 위에 종이를 깔고 칼로 곱게 다진다.

통잣은 전골 · 탕 · 신선로 등의 고명이나 차나 화채에 띄우고, 반 가른 비늘잣은 만두소나 편의 고명으로 쓰인다. 잣가루는 회나 적, 구절판 등에 쓰이고 초간장에도 넣는다.

• Pine Nuts

In Korean food, pine nuts have three types: 'tongjat' (whole grain without the pointed end), 'bineuljat' (a half of grain cut vertically), and pine nuts powder. Pine nuts powder is made by putting a piece of paper on a cutting board and pulverizing pine nuts with a knife. 'Tongjat' is used as garnish for stews, 'sinseonro' (hot pot of Korean royal cuisine), 'hwachae' (Korean fruit punch), and tea. 'Bineuljat' is used as garnish for dumpling stuffings and rice cakes. Pine nuts powder is used for raw seafoods, brochettes, 'gujeolpan' (Korean traditional dish with nine ingredients on a platter), and 'choganjang' (soy sauce mixed with vinegar).

• 호두

딱딱한 껍질 안의 알맹이를 반으로 갈라서 따뜻한 물에 담갔다가 꼬치로 속껍질을 벗긴다. 찜, 전골, 신선로의 고명으로 쓰인다.

• Walnuts

Walnuts are prepared by openning the hard shell, cutting the kernel in half, letting the kernels soak in warm water, and taking off the inner skin layer. It is used as garnish for steamed dishes, stews, and 'sinseonro' (hot pot of Korean royal cuisine).

• 대추

대추는 붉은색 고명으로 쓰이는데 주로 떡이나 과자에 많이 쓰인다. 돌려깎기해서 씨를 빼고 돌돌 말아 썰어서 꽃모양으로 만들거나 채 썰어서 쓴다.

• Jujube

Jujube is used as red garnish mostly for rice cakes and Korean traditional cookies. It is either julienned or cut into a flower shape by a rotational peeling.

• 밤

단단한 겉껍질과 속껍질까지 벗긴 후 찜에는 통째로 넣고, 떡에는 채로 썰거나 삶아서 체에 내려 고물로 쓰인다. 생률은 안주로 많이 쓰이고, 얇게 저며 썰어 보쌈김치 · 겨자채 · 냉채 등에 쓰인다.

• Chestnuts

Chestnuts are prepared by taking off the hard shell and the inner skin layer. For steamed foods, chestnuts are put in without cutting. For rice cakes, chestnuts are julienned. Peeled fresh chestnuts are often served with alchoholic beverages as a snack. Sliced fresh chestnuts are used for 'bossam kimchi' (kimchi for boiled pork belly), 'gyeoja-chae' (vegetables with mustard dressing), 'Naengchae' (cold vegetables), and other dishes.

● 식재료 Ingredients

• 두부

콩으로 만든 대표적인 식품으로 단백질이 풍부하고 식물성 지방이 들어 있다. 기원전 150년 중국의 회남왕 때부터 만들기 시작했다고 한다. 두부의 재료는 콩, 물, 간수(응고제)이며 만드는 방법은 콩을 불려 곱게 갈아 자루에 넣고 꼭 짠 두유를 모아서 솥에 끓인다. 여기에 간수를 넣으면 엉기는데 자루에 넣고 눌러서 굳히면 두부가 된다. 두부는 만들 때 가열시간과 응고제, 굳힐 때 누르는 힘에 따라 다양하게 만들 수 있다. 종류는 연두부, 순두부, 찌개두부, 부침두부 등 용도에 맞게 골라 쓴다.

• Dubu

Dubu is one of the most popular foods made from soybean. It is believed to be first made in 150 B.C. during the King Hoenam era in China. Dubu is rich in protein and vegetable fats, so it is good for health. Using soybeans, water, and brine as ingredients, Dubu is made by grinding finely soybeans soaked in water, putting the soybean powder into a sack, pressing it into soy milk, boiling the soy milk, making it condensed by adding brine, and finally solidifying it by putting it into a sack and pressing it. Depending on the heating time, coagulating agent, and the pressure under which it was solidified, dubu has four different types: 'yeondubu' (pudding dubu), 'sundubu' (soft dubu), dubu for stews, and dubu for frying.

• 북어

명태는 한국 사람들이 가장 즐겨 먹는 어류인데 명태를 말린 것을 북어라 한다. 북어 말리는 곳을 덕장이라 하는데, 12월 중순경 명태를 걸어 말리기 시작한다. 밤에는 얼고 낮에는 녹으면서 수분이 증발하고 살이 졸아들었다 부풀었다 하면서 결이 부드러워진다. 이 명태 중 상품을 황태라 하고 추위가 매서울수록 맛이 좋아진다. 술 마신 다음날 해장에도 좋은 북어에는 간에 활력을 불어넣어주고, 알코올을 분해하는 메티오닌, 타우린도 다량 들어 있다.

Bugeo (Dried Pollack)

Pollack is one of Korean's most favorite fish as food. The dried pollack is called 'bugeo' in Korean and the drying place is called 'deokjang'. From mid-December, pollack is hung up to be dried. The moisture in it evaporates as it gets frozen at night and melts in the daytime, and the texture of it gets softer as the skin shrinks and gets swollen. The colder the winter is, the better the taste is. Dried pollack with high quality is called 'hwangtae'. Dried pollack is good for relieving a hangover the day after drinking, and it revializes the liver with methionine and taurine, which help dissolve alcohol.

밀가루

밀의 구조는 배아, 내배유, 껍질의 세 부분으로 이루어져 있다. 내배유는 밀의 83~85%를 차지하고 이 부분을 제분하여 밀가루를 얻는다. 밀가루의 글루텐(밀 단백질) 함량에 따라 강력분, 중력분, 박력분으로 나뉜다. 강력분은 제빵용으로, 중력분은 면류, 만두피용으로 가정에서 가장 많이 쓰이며, 박력분은 제과용으로 쓰인다.

Flour

Wheat grain consists of three parts: gemmule, endosperm, and husk. The endosperm represents 83~85% of the wheat grain, which is milled to get flour. According to flour's content of gluten (wheat protein), flour falls into three types: strong flour, medium flour, and soft flour. Strong flour is used for bread. Medium flour is used for noodles and dumpling skins made at home. Soft flour is used for cookies.

멸치

멸치는 대부분 쪄서 말린 것을 사용한다. 멸치는 크기에 따라 국물용과 볶음용으로 나눈다. 국물용은 너무 큰 것보다는 중간 크기가 깊은 맛을 내고 작은 멸치들은 조림이나 볶음 등에 쓰인다.

• Myeolchi (Anchovy)

Anchovies used in Korean food are generally steamed and dried. Large or medium-sized anchovies are used for soup or broth (medium-sized anchovies are better for a deep taste), and small-sized anchovies are used for 'jorim' (food boiled down in soy sauce) and stir-fried dishes.

• 다시마

두툼하면서 하얀 분이 많이 묻은 것이 좋다. 성분 중에 감칠맛을 내는 지미성분이 있어 바로 만들어서 국물을 쓸 때 많이 쓰인다.

• Dasima (Kelp)

High-quality dasima is thick and covered with lots of white powder. Since dasima adds unami taste, it is widely used for broth.

• 김

김의 종류는 크게 돌김, 파래김, 재래김, 김밥김으로 나눈다. 육안으로 봐서 잡티가 적고 광택 나는 김이 좋다. 기름과 소금을 직접 발라서 굽거나 시판용 구이김으로 밥에 싸서 많이 먹는다.

• Gim (Laver)

Gim has three types: Dol-gim (stone laver), Parae-gim (green laver), Jaerae-gim (conventional laver), and Gimbap-gim (laver toasted to make gimbap). High-quality gim does not have blemishes and spots, and it is sleek and glossy. These days many people buy gim that is toasted and packaged, at a store rather than toast gim with oil and salt at home. People enjoy gim mostly by wrapping rice with it.

• 도라지

다년생초인 뿌리채소로서 주로 나물로 먹는다. 쓴맛을 제거할 때는 소금으로 주물러 씻거나 끓는 물에 데쳐낸다.

• Doraji (Balloon Flower Roots)

Doraji is a perennial herb and a root vegetable. Doraji is usually enjoyed as 'Namul' (Korean seasoned vegetables). To remove its bitter taste, it is rubbed with salt or blanched in boiling water.

• 고사리

나물이나 생선조림, 탕의 부재료로 많이 쓰인다. 생고사리는 반드시 삶아서 물에 우려낸 뒤에 먹어야 독성이 제거된다.

• Gosari (Fern Shoot)

Gosari is used for 'Namul' (Korean seasoned vegetables), 'saengseonjorim' (braised fish) and 'tang' (hot soup). Due to its toxicity, gosari should be boiled and brewed in water before eaten.

• 부추

생으로 무쳐 먹거나 데쳐서 무치기도 한다. 부추김치와 오이소박이의 소로도 쓰이며 여러 용도로 쓰인다. 제일 가는 영양부추, 중국요리에서 꽃빵과 같이 먹는 호부추, 그 밖의 요리에는 중간 정도 굵기를 쓴다. 간기능을 강화시키고 따뜻한 성질을 가지고 있다.

• Buchu (Korean Chives)

Raw or blanched buchu is served after being seasoned. Buchu is also used for various foods including 'buchu-kimchi' (kimchi made of buchu) and 'oi-sobagi' (cucumber kimchi). Buchu has several types. The 'yeongyang-buchu' is thin and short. The 'ho-buchu' is wide and usually served with Chinese flower bun. Buchu is known to improve the liver's function and make the body warm.

● 밑준비 Preparing Ingredients

● 마른 나물 불리기

마른 나물은 충분히 불린 후 삶아야 부드럽다. 불릴 때는 재료가 잠기도록 물을 부어 6시간 정도 불린 다음 찬물을 붓고 20~30분간 삶은 뒤 뚜껑을 덮어 그대로 식힌다. 아린 맛이 나는 나물은 여러 번 물을 갈아가며 1~2일 우려서 아린 맛을 제거한다. 고사리, 시래기(무청), 토란대, 말린 호박, 말린 고춧잎 등을 불릴 때 쓰는 방법이다.

● Soaking Dried Vegetables in Water

Dried vegetables should be soaked in water before boiling them to make them softer. Pour water till they are completely soaked and leave them alone for around 6 hours. After that, pour cold water and boil them for 20-30 minutes. Open the lid and let them cook down. For vegetables that taste bitter, replace the water several times and let them steep for a couple of days. It helps remove the bitter taste. It is used for cooking the following vegetables: bracken; dried radish leaves; taro stalks; dried green squash; dried chili leaves.

● 고기 핏물 제거법

· 볶음용 고기: 키친타월 등으로 가볍게 눌러 핏물을 제거하고 냉동고기의 경우 해동시켜서 핏물을 제거한다.

· 사골이나 갈비: 찬물에 담가 적어도 2시간 이상 놓아야 핏물이 빠진다. (소갈비의 경우 8시간 정도) 핏물을 덜 빼면 국물도 맑게 나오지 않고 누린내가 날 수 있다. 육수를 끓일 때 나오는 거품과 기름은 걷어내야 국물이 깨끗하고 맛도 좋다.

· 양지 육수: 고기 부위 중 육수를 만들었을 때 단맛이 난다. 핏물을 충분히 뺀 후 물이 끓을 때 고깃덩어리를 넣어서 끓이면 국물도 맑게 나오고, 고기는 건져서 썰거나 찢어서 양념한 후 건더기로 사용한다.

- How to remove blood from meat
  - Stir-fried meat : Press it gently with a paper towel to remove blood. For frozen meat, defrost it to remove blood.
  - Beef leg bone and rib : Let it soak in cold water for at least 2 hours. That's how its blood can be removed. (For beef rib, let it soak for around 8 hours.) If the blood is not removed sufficiently, the broth is not clear and there may be a bad odor. The bubbles and oil from boiling broth should be removed to make the texture clear and to make it taste better.
  - Brisket broth : It tastes sweet, unlike other parts of meat. Remove all blood and put in a chunk of meat when the water starts to boil. Then the broth is clear and tastes better. Pick up the meat and chop it or rip it. Season it and use it as the stock.

- 멸치 다시마 육수 만들기

멸치는 배 쪽의 내장을 제거하고 마른 팬에 볶아 비린 맛을 제거한다. 다시마는 면보에 물을 적셔 꽉 짠 후, 겉에 묻은 하얀 분을 닦는다. 물을 붓고 찬물부터 멸치와 다시마를 넣고 끓이다가 끓어오르면 불을 줄여 10분 정도 끓인 다음 체에 걸러낸다.

- Making Myeolchi Dasima Yuksu (Dried anchovy and Kelp broth)

Remove the intestines of anchovies and stir-fry them to remove fish odor. For kelp, remove the white powder with wet cloth. Put the stir-fried anchovies and prepared kelp in a pot and boil them. When the water starts boiling, reduce the heat, boil for another 10 minutes, and strain the broth through a sieve.

- 조개 해감

찬물에 약간의 소금을 타서 녹인 후 조개를 담가뒀다가 손으로 비벼 씻는다. 봉지에 들어 있는 것을 사면 해감이 잘 되어 있다.

- Cleaning Jogae (Clams)

Put the small amount of salt in a bowl with cold water. Let the clams sit in the bowl for a while and rub them with hands. If you buy the one from supermarket, it should be clean enough to cook it right away.

- 새우 내장 제거하는 법

새우나 대하 몸통의 2~3마디째 되는 곳에 이쑤시개나 가는 꼬치로 찔러 넣어 당기면 검은색 가는 긴 줄의 내장이 따라 나온다.

- Removing Intestines of Saewoo (shrimps/prawns)

Stab the second or third joints of shrimps/prawns with a sharp toothpick or a skewer. Pull it out to remove black-colored thick and long intestines.

● 썰기

써는 식품의 종류 및 용도에 따라 칼의 사용부분과 동작의 방향이 정해진다.

● Cutting

Depending on the type and use of food, which part of a knife and how it is used is determined.

기본 썰기
Basic Cutting

- 통썰기

모양이 둥글고 긴 재료인 오이, 당근, 연근 등을 통째로 써는 방법

- Cutting in Chunks (Chopping)

It is used for cutting round and long materials such as cucumber, carrot, and lotus root.

- 반달썰기

통으로 썰기에 너무 큰 재료인 무, 고구마, 감자 등을 길이로 반 가른 후 반달 모양으로 썬다.

- Half-moon Shape Cutting

It is used for cutting radish, potato, and sweet potato. Cut them in half into a half-moon shape.

• 은행잎 썰기

재료를 십자로 4등분한 다음 고르게 은행잎 모양으로 썬 것. 감자, 무, 당근 등을 조림하거나 찌개에 넣을 때 쓰인다.

• Gingko Leaf-shape Cutting (Fermière)

Divide a material into four parts in the form of a cross and cut it into a gingko leaf shape with equal pieces. It is used for boiling potato, radish, and carrot, or putting them in a stew.

• 편썰기

마늘이나 생강을 손질한 후 마늘은 길이대로 얇게 썰고 생강은 모양대로 얄팍하게 썬다. 채를 잘 썰려면 얇고 균일하게 편으로 써는 과정을 거쳐야 한다.

• Cutting into Thin Pieces (Slicing)

Cut trimmed garlics or gingers into thin pieces. It is required for shredding a vegetable. Each piece should be thin and equal.

• 어슷썰기

가늘고 긴 재료인 오이, 당근, 고추, 파 등을 적당한 두께로 사선으로 써는 방법

• Diagonal Cutting (Bias-cutting)

Cutting long, thin vegetables such as cucumber, carrot, pepper, and green onion diagonally in proper thickness.

• 골패썰기 · 나박썰기

토막 낸 재료를 네모지게 잘라내고 직사각형으로 얇게 썬 것은 골패썰기, 사각형으로 썰면 나박썰기이다.

• Rectangular-shape Cutting, Square-shape Cutting (Lozenge)

Cut vegetables into thin rectangular pieces or square pieces depending on the item.

• 깍둑썰기

무, 감자, 두부 등을 사각으로 막대썰기한 다음 다시 주사위 모양으로 썬다. 깍두기, 조림, 찌개 등에 흔히 쓰인다.

• Cube Cutting (Large Dice)

Cut radish, potato, or dubu into sticks and cut them into dice shapes. It is used for radish Kimchi, braised food, and stew.

• 채 썰기

무, 양파, 당근, 오이 등을 얄팍하게 썬 것을 비스듬히 포개 놓고 가늘게 썬 것. 생채나 잡채 등에 주로 쓰이는 썰기

• Shredding (Julienne)

First, cut radish, onion, carrot, and cucumber into thin slices. Second, put them together in a pile and cut them into thin pieces. It is often used for raw vegetables and glass noodle garnish.

• 다지기

채 썬 것을 가지런히 놓고 직각으로 잘게 써는 방법. 파, 마늘, 생강, 양파 등의 양념을 만드는 데 주로 쓰인다.

• Chopping into Thin Pieces (Mincing)

Put the shredded pieces together and chop them into thin pieces. It is used for making spice mixes with green onion, garlic, ginger, or onion.

• 마구썰기

가늘고 긴 재료인 오이나 당근, 우엉 등을 한 손으로 돌려가며 한입 크기로 각지게 써는 방법

• Rolling Cutting (Oblique)

Cut long and thin vegetables such as cucumber, carrot, and burdock into bite-size pieces fast.

• 돌려깎기

오이나 호박, 무, 당근 등을 얄팍하게 포 뜨듯 돌려 깎아 채로 썰 때 많이 쓰는 방법. 재료의 돌려깎을 방향으로 칼을 사선으로 꽂은 후 재료는 칼날 방향으로, 칼은 그 반대 방향으로 균일하고 얄팍하게 재료를 깎는 방법

• Rotation Cutting (Turned)

Cut cucumber, green squash, radish, and carrot in a ring and slice them into thin pieces. It is often used for shredding. Put a knife in the vegetable diagonally in the direction of cutting and cut it into thin slices.

• 송송 썰기

고추, 대파 등을 모양대로 둥글게 썰거나, 김치를 길이로 2~3등분한 후 원하는
크기대로 잘게 썰 때 쓰인다.

• Chopping into Chunks (Small Dice)

Cut pepper or large green onion into round
pieces. It is also used for cutting Kimchi after
dividing it into two or three parts.

불고기 *Bulgogi* / 오이송송이 *Oi-songsongi* / 콩나물국 *Kongnamul-guk* / 애호박전 *Aehobak-jeon* / 육원전 *Yugwon-jeon* / 생선전 *Saengseon-jeon* / 된장찌개 *Doenjang-jjigae* / 순두부 *Sundubu* / 돼지고기김치찜 *Doejigogi-kimchi-jjim* / 호박나물 *Hobak-namul* / 규아상 *Gyuasang* / 죽순겨자채 *Juksun-gyeoja-chae* / 땅콩호두장과 *Ttangkong-hodu-janggwa* / 비빔밥 *Bibimbap* / 고사리나물 *Gosari-namul* / 도라지나물 *Doraji-namul* / 시금치나물 *Sigeumchi-namul* / 뭇국 *Mu-guk* / 닭산적 *Dak-sanjeok* / 김밥 *Gimbap* / 달걀북엇국 *Dalgyal-bugeo-guk* / 잡채 *Japchae* / 배추겉절이 *Baechu-geotjeori* / 미역국 *Miyeok-guk* / 삼계탕 *Samgye-tang* / 양배추김치 *Yangbaechu-kimchi* 깻잎찜 *Kkaennip-jjim* / 마늘장아찌 *Maneul-jangajji* / 삼겹살과 쌈 상 *Samgyeopsal & Ssam-sang* / 제육구이 *Jeyuk-gui* / 조개탕 *Jogae-tang*

봄 / 여름

불고기, 콩나물국, 오이송송이

부모님 따라나섰던 나들이 길에 거북이 등처럼 생긴 판 위에서 익어가던
불고기와 남겨진 들척지근한 국물에 밥 한 수저 비벼 먹던 그 맛

Bulgogi, Kongnamul-guk,
Oi-songsongi

I yearn for the taste of food that I had together with my parents at picnics,
enjoying a spoonful of rice with Bulgogi and the left over sauce, cooked on the
turtle-shell looking plate.

불고기

불고기는 너비아니에서 기원했다. 너붓너붓 썰어 숯불에 굽는 옛 방식의 너비아니와 다르게
요즈음의 불고기는 얇게 썰어 양념에 재운 방식이다. 불고기에 들어가는 배즙은 소화를 돕고
고기를 연하게 한다. 하나의 주음식으로 손색이 없다.

Bulgogi_ *Stir-fried or Grilled Beef*

Bulgogi originated from Neobiani (너비아니). Different from old-fashioned Neobiani which is a thick
slice grill, modern Bulgogi is thinly sliced and cooked after being marinated in Bulgogi sauce. Pear juice
helps digestion and meat tenderization. This dish is a great main dish.

재료

주재료
쇠고기 600g (등심, 얇게 한입 크기로 썬 것)

부재료
양파 240g (길이로 썰기) • 대파 50g (어슷썰기) • 배즙 1C

양념
간장 4T • 설탕 2T • 후추 약간 • 다진 마늘 1½T • 다진 파 1T • 참기름 2T • 깨소금 1t

고명
통깨 약간

만드는 법

1. 연육하기
　-핏물 제거한 고기를 배즙에 15분 정도 담가둔다.
2. 양념 만들기
　-큰 그릇에 양념을 모두 넣고 섞는다.
3. 고기 재우기
　-No. 1 고기와 No. 2 양념을 잘 섞는다.
　-채소를 넣어준다.
　-15분간 재워둔다.
4. 고기 볶기/ 굽기
　-예열된 팬에 재워둔 고기를 넣고 볶는다.
　-혹은 숯불이나 그릴을 이용해 잘 구워준다.
5. 담아내기 및 고명 올리기
　-그릇에 옮겨 담고 통깨를 살짝 뿌려준다.

Ingredients

Main ingredients
Beef 600g (Sirloin, thinly sliced, bite-size)

Sub-ingredients
Onion 240g (julienne) • Green Onion 50g (bias-cut) • Korean Pear Juice 1C

Seasonings
Soy Sauce 4T • Sugar 2T • Black Pepper pinch • Garlic 1½T (minced) • Green Onion 1T (finely chopped) • Sesame Oil 2T • Sesame Seeds 1t (ground)

Garnish
Sesame Seeds pinch

Steps

1. Tenderizing meat
　-Let blooded beef marinate in pear juice for 15 minutes.
2. Making sauce
　-Mix all seasonings in a big bowl.
3. Marinating
　-Mix no. 1 with no. 2.
　-Add vegetables.
　-Marinate for 15 minutes.
4. Stir-frying/ grilling
　-Stir-fry the marinade on a preheated pan.
　-Cook thoroughly using grill or broiler.
5. Plating and garnishing
　-Transfer to a plate with a pinch of sesame seeds on top.

귀띔
Tips

1. 고기의 누린내를 없애기 위해 핏물을 충분히 빼준다.
2. 팬에 볶는 대신, 숯불이나 그릴을 사용하면 식감과 풍미를 더할 수 있다.
3. 불고기는 쌈과 잘 어울린다.

1. The meat needs to be blooded enough to remove the gamy smell.
2. Instead of stir-frying, grilling or broiling enhances the texture and taste of meat.
3. Bulgogi goes well with Ssam which is wrapped with assorted vegetables.

오이송송이

궁중에서는 깍두기를 '송송이'라 하는데 오이송송이는 오이를 깍두기처럼 썰어 담근다 하여
붙여진 이름이다.

Oi-songsongi_*Cucumber-kimchi*

In Joseon Dynasty, Kkakdugi was called 'Songsongi.' In Oi-songsongi, cucumber (Oi in Korean) is diced
like Kkakdugi.

재료

주재료
오이 5개 · 소금 1T · 물 1C

부재료
실파 20g (3cm 길이로 썰기)

양념
고춧가루 2T · 다진 마늘 1T · 다진 생강 1t · 설탕 1t · 새
우젓 1½T (곱게 다지기) · 소금 약간

만드는 법

1. 오이 손질하기
- 오이는 소금으로 문질러 깨끗이 씻는다.
- 오이는 통으로 길게 4등분하여 씨를 제거한 뒤 3cm 길이로 썬다.

2. 절이기
- 오이 썬 것에 소금물을 부어 10분간 절인다.
- 절여진 오이는 채반에 건져 물기를 뺀다.

3. 버무리기
- 물기 뺀 오이에 고춧가루를 먼저 넣어 버무려둔다.
- 모든 양념을 넣고 잘 버무려준다.
- 실파를 넣고 살짝 섞어준다.

4. 담아내기
- 그릇에 담아낸다.

Ingredients

Main ingredients
Cucumber 5ea · Salt 1T · Water 1C

Sub-ingredients
Small Green Onion 20g (cut 3cm length)

Seasonings
Red Chili Powder 2T · Garlic 1T (minced) · Ginger 1t
(minced) · Sugar 1t · Salted Shrimp 1½T (finely chopped) · Salt
pinch

Steps

1. Preparation
- Rub cucumber with salt and rinse with water.
- Slit cucumber into four equal pieces, remove the seeds and cut them in 3cm length.

2. Brining
- Soak cucumber in salt water for 10 minutes.
- Strain off brine from soaked cucumber.

3. Seasoning
- Add red chili powder to salted cucumber.
- Mix all the seasoning ingredients with the cucumber.
- Put small green onion in the mixture.

4. Serving
- Transfer to a plate.

콩나물국

콩나물국은 한국인들이 매우 좋아하는 맑은국이다. 아스파라긴산이 풍부한 콩나물은 특히 숙취해소에 탁월하다.

Kongnamul-guk_ *Soybean Sprout Soup*

Kongnamul-guk is a favorite clear soup for most Koreans. Especially, it is good for hangovers since soybean sprout contains aspartic acid.

재료

주재료
콩나물 240g (다듬기)

부재료
멸치 25g (내장 제거) • 다시마 20g (약 10×10cm) • 물 8C
(≒1.6L) • 대파 50g (어슷썰기) • 다진 마늘 1/2T • 풋고추
1개 (어슷썰기)

양념
소금 2t

만드는 법

1. 국물내기
- 비린내를 없애기 위해 멸치를 팬에 볶는다.
- 다시마와 물을 넣고 10분간 끓인다.
- 국물이 우러난 멸치와 다시마는 건져낸다.

2. 콩나물 넣어 끓이기
- 콩나물과 소금을 넣고 끓인다.
- 마늘과 풋고추를 넣는다.
- 대파를 넣고 불을 끈다.

3. 담아내기
- 국그릇에 옮겨 담는다.

Ingredients

Main ingredients
Soybean Sprout 240g (trimmed)

Sub-ingredients
Dried Anchovy 25g (drawn) • Kelp 20g (10×10cm) • Water 8C (≒
1.6L) • Green Onion 50g (bias-cut) • Garlic 1/2T (minced) • Green
Chili 1ea (bias-cut)

Seasonings
Salt 2t

Steps

1. Making stock
- Toast anchovies in a sauce pan to remove fishy smell.
- Add kelp and water. Bring it to a boil for 10 min.
- Take anchovies and kelp out of the stock.

2. Adding soybean sprout
- Add soybean sprout and salt. Bring it to a boil.
- Add garlic and green chili.
- Add green onion and turn off heat.

3. Plating
- Transfer to a soup bowl.

귀띔
Tips

1. 멸치를 기름 없는 팬에 살짝 볶아주면 수분이 날아갈 뿐 아니라 비린내도 없앨 수 있다.
2. 콩나물을 넣고 끓일 때 뚜껑을 닫아주어야 비린내가 나지 않는다.
3. 콩나물국은 맛이 깔끔하고 순하기 때문에 강한 맛의 음식과 잘 어울린다. 때로는 매콤한 맛을 더하기 위해
   고춧가루를 넣기도 한다.

1. Toasting dried anchovy eliminates its water and smell.
2. Boiling sprouts with a lid on prevents the soup from having uncooked bean smell.
3. Kongnamul-guk goes well with dishes that have complex and strong seasonings, since it is clear and
   mild. To enhance its spiciness, chili powder is often added.

생선전, 애호박전, 육원전, 된장찌개
집으로 돌아오는 골목 어귀에 된장 끓는 소리와 전 지지는 냄새는
발걸음을 재촉하게 한다.

Saengseon-jeon, Aehobak-jeon,
Yugwon-jeon, Doenjang-jjigae
The sound of Doenjang-jjigae boiling and the aroma of Jeon frying on the cover
of an iron pot, these make my steps hasten once I see them on my way back
home.

애호박전

'전'은 한국의 전통적인 조리법이다.
애호박전은 애호박을 둥글게 썰어 밀가루와 달걀을 입혀 지지는 전이다.

Aehobak-jeon_ *Pan-fried Green Squash*

Jeon is one of the traditional Korean cooking methods.
Aehobak-jeon is pan-fried green squash coated in flour and beaten egg.

재료

주재료
애호박 1개 (0.5cm 두께로 둥글게 썰기)

부재료
밀가루 1C · 달걀 2개 (풀기) · 식용유

양념
소금 약간

초간장
간장 1T · 식초 1/2T · 설탕 1/2T · 물 1T · 잣가루 약간

만드는 법

1. 양념
- 썰어놓은 애호박에 소금을 뿌려둔다. (5분 정도)
- 키친타월이나 면보로 여분의 물기를 제거한다.

2. 밀가루 입히기
- 애호박에 밀가루를 고르게 입힌다.
- 여분의 밀가루를 털어낸다.
- 풀어놓은 달걀을 입힌다.

3. 부치기
- 예열한 팬에 기름을 충분히 두른다.
- 양면을 완전히 익힌다.

4. 담아내기
- 접시에 담는다. 초간장과 곁들여 낸다.

Ingredients

Main ingredients
Green Squash 1ea (0.5cm sliced, round)

Sub-ingredients
Flour 1C · Egg 2ea (beaten) · Cooking Oil

Seasonings
Salt pinch

Dip
Soy Sauce 1T · Vinegar 1/2T · Sugar 1/2T · Water 1T · Pine Nuts pinch (ground)

Steps

1. Seasoning
- Sprinkle salt over squash slices (set aside about 5 minutes).
- Set aside.
- Tap them to dry out using cotton cloth.

2. Dredging
- Coat squash slices with flour evenly.
- Shake off excess flour.
- Coat them with beaten egg.

3. Pan-frying
- Preheat a pan with enough oil.
- Cook each side throughly.

4. Plating
- Transfer to a plate. Serve it with dipping sauce.

외국인도 빠져드는 한국밥상

59

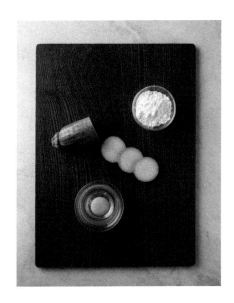

육원전

전은 생선이나 고기, 채소 등을 얇게 썰거나 다져서 지지는 조리법이다.
육원전은 고기와 두부를 섞어 둥글납작하게 만들어 지지는 전이다.

Yugwon-jeon

__ Pan-fried Mixture of Beef and Dubu

Jeon is a Korean cooking method of panfrying made of vegetables, meat, or fish.
Yugwon-jeon is a pan-fried mixture of beef and dubu, coated with flour and beaten egg.

재료

주재료
다진 쇠고기 100g · 두부 50g

부재료
밀가루 1/4C · 달걀 1개 (풀기) · 식용유

양념
소금 1/2t · 설탕 1/2t · 다진 파 1t · 다진 마늘 1/2t · 참기름 1t · 후춧가루 약간

초간장
간장 1T · 식초 1/2T · 설탕 1/2T · 물 1T · 잣가루 약간

만드는 법

1. 두부 으깨기
- 칼등으로 두부를 으깬다.
- 면보로 여분의 물기를 짜낸다.

2. 양념하기
- 으깬 두부와 다진 쇠고기를 볼에 담고 양념을 섞는다.

3. 모양 만들기
- 한 숟갈씩 떼어낸다.
- 작은 공처럼 만든다.
- 가운데를 눌러 둥글납작하게 만든다.

4. 밀가루 입히기
- 밀가루를 골고루 입힌다.
- 여분의 밀가루를 털어낸다.
- 달걀을 입힌다.

5. 부치기
- 기름을 충분히 두르고 팬을 예열한다.
- 양면을 완전히 익힌다.

6. 담아내기
- 접시에 담는다.

Ingredients

Main ingredients
Beef 100g (minced) · Dubu 50g

Sub-ingredients
Flour 1/4C · Egg 1ea (beaten) · Cooking Oil

Seasonings
Salt 1/2t · Sugar 1/2t · Green Onion 1t (finely chopped) · Garlic 1/2t (minced) · Sesame Oil 1t · Black Pepper pinch

Dip
Soy Sauce 1T · Vinegar 1/2T · Sugar 1/2T · Water 1T · Pine Nuts pinch (ground)

Steps

1. Mashing dubu
- Mash it using the back of knife.
- Squeeze dubu using cotton cloth to remove excess water.

2. Seasoning
- Mix no. 1 and minced beef with all the seasonings in a bowl.

3. Shaping
- Divide no. 2 into tablespoonful portions.
- Make them into balls.
- Press them on the center to flatten.

4. Dredging
- Coat them with flour evenly.
- Shake off excess flour.
- Coat them with beaten egg.

5. Pan-frying
- Preheat a pan with enough oil.
- Cook each side throughly.

6. Plating
- Transfer to a plate.

61

생선전

전은 기념일이나 잔치음식으로 만들어 먹는다.
전은 맛이 좋을 뿐만 아니라 한입 크기로 만들어 먹기도 쉬워서 어린 아이부터 노인에 이르기까지
여전히 사랑받고 있다.
생선전은 주로 흰살 생선을 이용하여 지지는 전이다.

Saengseon-jeon_ *Pan-fried Fish Fillet*

Jeon is prepared for memorial services or festivals. Jeon is not only tasty but also easy to eat as a bite size, so even today it remains popular among both the young and old.

재료

주재료
흰살 생선 150g

부재료
밀가루 1/2C · 달걀 1개 (풀기) · 식용유

양념
흰 후춧가루 약간 · 소금 약간 · 생강즙 1/4t

초간장
간장 1T · 식초 1/2T · 설탕 1/2T · 물 1T · 잣가루 약간

만드는 법

1. 생선 저미기
 - 생선살을 0.7cm 두께로 포를 뜬다.

2. 양념하기
 - 소금과 후춧가루를 생선살에 뿌린다.
 - 풀어놓은 달걀에 소금 간을 한다.

3. 밀가루 입히기
 - 생선살에 밀가루를 고르게 입힌다.
 - 여분의 밀가루를 털어낸다.
 - 풀어놓은 달걀을 입힌다.

4. 부치기
 - 예열한 팬에 기름을 충분히 두른다.
 - 양면을 완전히 익힌다.

5. 담아내기
 - 접시에 담는다.

Ingredients

Main ingredients
White Fish Fillet 150g

Sub-ingredients
Flour 1/2C · Egg 1ea (beaten) · Cooking Oil

Seasonings
White Pepper pinch · Salt pinch · Ginger Juice 1/4t

Dip
Soy Sauce 1T · Vinegar 1/2T · Sugar 1/2T · Water 1T · Pine Nuts pinch (ground)

Steps

1. Filleting
 - Fillet fish into 0.7cm thick slices.

2. Seasoning
 - Sprinkle salt and white pepper and add ginger juice on fillets.
 - Season beaten egg with salt.

3. Dredging
 - Coat fillets with flour evenly.
 - Shake off excess flour.
 - Coat fillet with beaten egg.

4. Pan-frying
 - Preheat a pan with enough oil.
 - Cook each side throughly.

5. Plating
 - Transfer to a plate.

귀띔 흰살 생선으로는 광어, 대구, 도미, 가자미, 명태를 사용한다.
Tips Flounder, codfish, seabream, plaice and pollack can be used for white fillet fish.

된장찌개

된장은 콩을 잘 익히고 발효시켜 만든 전통적인 양념 중 하나이다.
된장찌개는 된장을 풀어 여러 가지 채소와 육류, 해물 등을 넣고 함께 끓인 음식이다.

Doenjang-jjigae_ *Soybean Paste Stew*

Doenjang, soybean paste is one of the traditional Korean seasonings; it is made as a thick paste from ripened and fermented soybeans.

재료

주재료
쇠고기 100g (양지머리, 잘게 썰기) • 물 3C • 된장 3T

부재료
두부 150g (3×2.5×1cm 크기로 썰기) • 애호박 1/4개 (반달 썰기) • 건표고 3개 (불리기, 4조각내기) • 풋고추 1개 (어슷썰기) • 홍고추 1개 (어슷썰기) • 대파 1/2개 (어슷썰기)

양념
쇠고기 밑양념 • 국간장 1t • 다진 파 2t • 다진 마늘 1t • 후춧가루 약간

만드는 법

1. 쇠고기 양념하기
- 쇠고기에 양념을 넣고 밑간을 한다.

2. 된장육수 만들기
- 물에 된장을 풀어 끓으면 양념한 쇠고기를 넣고 끓인다.

3. 채소 넣기
- 애호박과 버섯을 넣고 끓인다.
- 두부를 넣고 끓이는 동안 위에 뜬 거품을 걷어낸다.
- 고추, 대파를 넣는다.
- 한소끔 더 끓인다.

4. 담아내기
- 찌개용 그릇에 담아낸다.

Ingredients

Main ingredients
Beef 100g (brisket, small bite-size cut) • Water 3C • Doenjang 3T

Sub-ingredients
Dubu 150g (3×2.5×1cm sliced) • Green Squash 1/4ea (halfmoon sliced) • Dried Pyogo Mushroom 3ea (soaked, quartered) • Green Chili 1ea (bias-cut) • Red Chili 1ea (bias-cut) • Green Onion 1/2ea (bias-cut)

Seasonings
For beef • Gukganjang 1t • Green Onion 2t (finely chopped) • Garlic 1t (minced) • Black Pepper pinch

Steps

1. Marinating beef
- Mix all the marinade seasonings.
- Mix beef with beef seasoning in a bowl. Set aside.

2. Making soup base
- In water, add soybean paste and stir well to dissolve.
- Add no. 1 and wait to be cooked.

3. Adding vegetables
- Add green squash and mushroom. Bring it to boil.
- Add dubu. Scum off, while boiling.
- Add chili, green onion.
- Bring it to a boil.

4. Plating
- Transfer to a bowl.

귀띔 쇠고기 대신 다시마 또는 멸치 육수를 써도 된다.
Tips Instead of beef, kelp or anchovy broth can be used.

돼지고기김치찜, 순두부, 호박나물

왼손으론 불린 콩을 옆으로 새지 않게 한 수저씩 떠 넣고, 오른손은 뻗어
맷돌자루를 돌려가며 갈아 공들여 끓여 만든 콩물에 간수를 부어 몽글몽글해지면,
자박지(질그릇으로 만든 넓은 그릇)에 그득 떠주시던 어머니

Doejigogi-kimchi-jjim, Sundubu, Hobak-namul

My mom used to make dubu at home. With her left hand, she scooped up a spoonful of
soaked bean and fed it to a mill. With her right hand, she turned the handle to grind
the beans and pour the sea water to make clotty tofu. Once it was made, she gave me
a bowlful of newly made tofu in a special dish called Jabakji.

순두부

순두부는 콩으로 만든 완벽한 자연식 중 하나이다.
한국 사람들은 순두부, 두부(모두부: 순두부를 눌러 물기를 제거한 것), 비지(두부를 만들고 난
찌꺼기)를 매우 좋아한다. 순두부는 부드럽고 풍미가 풍성하며 소화하기 쉬워 유아나 아이들,
노인, 그리고 환자에게 좋은 음식이다.

Sundubu_ *Soft Dubu*

Sundubu is one of the super foods made from soybeans. Koreans love Sundubu, Dubu (or Modubu,
drained & pressed sundubu) and Biji (leftover of soaked & squeezed soybeans) very much. Sundubu
is silky, nutty and easy to digest. It is especially good for toddlers, children, the elderly and medical
patients.

재료

주재료
마른 메주콩 3C (찬물 9C, 충분히 불리기, 12시간 이상) · 따뜻한 물 20~24C (50~60℃, 68~75℉)

부재료
들기름 1/2~1T · 간수 3~4T

양념
소금 1T · 다진 파 1T · 다진 마늘 1T · 깨소금 1T · 참기름 1T · 간장 3T · 굵은 고춧가루 1t · 다진 청 · 홍고추 각 1t씩

고명
청 · 홍고추 각 1/2개씩 (씨 빼기, 어슷썰기)

만드는 법

1. 불린 콩 갈기
- 분쇄기(혹은 믹서기)에 콩을 넣는다.
- 콩과 물을 분쇄기에 넣고 매우 곱게 갈아준다.

2. 짜기
- 간 콩을 면 주머니에 넣고 따뜻한 물을 3~4번에 나누어 부어 진한 콩물을 짜낸다.

3. 거품 제거하기
- 들기름을 콩 국물에 고르게 뿌려 거품을 제거한다.

4. 끓이기와 굳히기
- 콩 국물을 끓인다. 콩물이 눌어붙지 않도록 끓을 때까지 계속 저어준다. 콩 국물이 끓으면 바로 불을 끈다.
- 콩 국물 표면에 막이 생기면 간수를 고르게 뿌린다.
- 간수가 고르게 퍼지도록 주걱을 세워 살살 갈라준다.

5. 양념장 만들기
- 모든 양념을 골고루 섞는다.

6. 담아내기
- 깊고 둥근 대접에 순두부를 담는다.
- 청 · 홍고추를 고명으로 얹는다.
- 뜨거울 때 양념장과 같이 낸다.

Ingredients

Main ingredients
Dry Soybeans 3C (water 9C fully soaked over 12 hours)
Warm Water 20~24C (50~60℃, 68~75℉)

Sub-ingredients
Perilla Oil 1/2~1T · Bittern 3~4T

Seasonings
Salt 1T · Green Onion 1T (finely chopped) · Garlic 1T (minced) · Sesame Seeds 1T (toasted & ground) · Sesame Oil 1T · Soy Sauce 3T · Chili Powder 1t · Green Chili 1t (finely chopped) · Red Chili 1t (finely chopped)

Garnish
Green Chili 1/2ea (seeded, bias-cut) · Red Chili 1/2ea (seeded, bias-cut)

Steps

1. Grinding soybeans
- Grind the soaked beans with soaking water.
- Grind until it becomes soft paste.

2. Squeezing
- Squeeze the bean juice with warm water.
- Repeat this steps 3~4 times.

3. Removing bubbles
- Sprinkle perilla oil evenly into the juice to remove bubbles.

4. Boiling & Curdling
- Boil the juice. Keep stirring to prevent it from sticking. When starting to boil, turn off the heat.
- Sprinkle bittern evenly when white film shows on the surface of the bean juice.
- Stir very gently to mix bittern.

5. Making dressing
- Mix all seasonings.

6. Plating
- Transfer Sundubu into a deep round bowl.
- Garnish with green and red chili.
- Serve hot with dressing.

69

귀띔 Tips

1. 두부는 해콩으로 만드는 것이 좋다.
2. 콩을 불릴 때 콩이 쉬지 않도록 서늘한 곳에 보관해야 한다.
3. 순두부는 원래 따뜻할 때 양념장만 넣어 먹던 음식이었다. 요즘은 조갯살이나 굴 또는 돼지고기를 넣고 맵게 끓이기도 한다.

1. To make best quality Dubu, should use the new soybean.
2. When the beans get soaked, it needs to stay in cool area to prevent beans from spoiling.
3. Originally, Sundubu is served hot only with dressing. Todays, people make spicy Sundubu with clam, oyster or pork.

돼지고기김치찜

한국 사람들은 삼겹살과 김치를 좋아한다.
돼지고기김치찜은 돼지고기와 김치를 가지고 만드는 대중적인 음식으로 맵고 풍미가 있다.

Doejigogi-kimchi-jjim_ *Braised Pork and Kimchi*

Korean love pork belly (Samgyopsal) and Kimchi. Doejigogikimchi-jjim is one of the popular dishes which made with pork belly and Kimchi. It is spicy and full of flavory.

재료

주재료
돼지고기 400g (삼겹살, 얇게 썬 것, 한입 크기로 자르기) · 김치 800g (속 털어내기, 한입 크기로 자르기)

부재료
가래떡 300g (5cm 길이, 4등분하기) · 대파 50g (어슷썰기) · 양파 70g (길이로 썰기) · 풋고추 2개 (씨 빼기, 어슷썰기) · 홍고추 1개 (씨 빼기, 어슷썰기) · 물 1/2C · 식용유 2T · 참기름 1/2T

양념
돼지고기 밑양념 · 간장 2T · 다진 마늘 1T · 생강즙 1/2T · 고춧가루 1T · 설탕 1T · 후추 약간
김치 밑양념 · 다진 마늘 1/2T · 설탕 1T
가래떡 밑양념 · 참기름 1/2T
고명 · 잣 1T

만드는 법

1. 양념 만들기
-준비한 돼지고기 밑양념을 골고루 섞는다.
-준비한 김치 밑양념을 골고루 섞는다.

2. 양념에 재우기
-돼지고기 밑양념을 돼지고기에 넣고 버무린다.
-김치 밑양념을 김치에 넣고 버무린다.
-가래떡 밑양념을 가래떡에 넣고 버무린다.

3. 볶기
-예열한 팬에 식용유를 넣고 김치를 볶는다.
-돼지고기를 넣고 완전히 익을 때까지 볶는다.
-양파를 넣고 골고루 뒤적인다.

4. 끓이기
-물과 가래떡을 넣고 골고루 섞어가며 볶는다.
-팬 뚜껑을 덮고 중불로 조린다.
-대파, 청·홍고추를 넣고 골고루 섞는다.
-마지막에 참기름을 넣고 뒤적인다.

5. 담아내기
-잣을 넣고 골고루 섞는다.
-뜨거울 때 접시에 담아낸다.

Ingredients

Main ingredients
Pork Belly 400g (thin sliced, bite-size-cut) · Kimchi 800g (trimmed, bite-size-cut)

Sub-ingredients
Garae-tteok 300g (stick rice cake, bite size-cut) · Green Onion 50g (bias-cut) · Onion 70g (julienne) · Green Chili 2ea (seeded, bias-cut) · Red Chili 1ea (seeded, bias-cut) · Water 1/2C · Cooking Oil 2T · Sesame Oil 1/2T

Seasonings
For pork · Soy Sauce 2T · Garlic 1T (minced) · Ginger Juice 1/2T · Red Chili Powder 1T · Sugar 1T · Black Pepper pinch
For kimchi · Garlic 1/2T (minced) · Sugar 1T
For garae-tteok · Sesame Oil 1/2T
Garnish · Pine Nuts 1T

Steps

1. Making marinade sauce
-Mix all seasoning for pork.
-Mix all seasoning for kimchi.

2. Marinating
-Add pork marinade sauce into pork. Mix well.
-Add kimchi marinade sauce into kimchi. Mix well.
-Add Garae-tteok marinade sauce into Garae-tteok. Mix well.

3. Stir-frying
-Preheat a pan and add cooking oil. Stir-fry kimchi throughly.
-Add pork and stir-fry until it becomes well-done.
-Add onion and stir well.

4. Simmering
-Add water and Garae-tteok. Stir well.
-Cover the pan, cook throughly.
-Add green onion, green chili and red chili. Mix well.
-Add sesame oil and mix well.

5. Plating
-Sprinkle pine nuts.
-Serve hot. Transfer to a plate.

71

귀띔
Tips
1. 생강즙 대신 적포도주 혹은 청주 1T를 사용해도 된다.
2. 굳은 가래떡은 데쳐서 사용하는 것이 좋다.

1. Ginger juice can be replaced with 1 tablespoon of red wine or rice wine (Cheong-ju).
2. If Garae-tteok is too hard, blanch it before cooking.

호박나물

호박나물은 한식상차림에 주요 찬으로 올리는 음식 중 하나이다. 애호박을 반달썰기한 뒤 양념하여 볶은 것이다.

Hobak-namul_ *Stir-fried Green Squash*

Hobak-namul is one of the most popular side dishes in Korean food.
It is a stir-fried squashes seasoned and sliced in a half-moon shape.

재료

주재료
애호박 400g (반달모양, 0.5cm 두께로 썰기)

부재료
물 1/2C · 소금 1T · 식용유 2T

양념
다진 파 1T · 다진 마늘 1t · 깨소금 1/2T

고명
홍고추 10g (어슷썰기)

만드는 법

1. 소금에 절이기
- 소금물을 만든다.
- 애호박을 넣고 가볍게 섞는다.
- 10분 정도 지난 뒤 물기를 제거한다.

2. 볶기
- 예열된 팬에 식용유를 넣는다.
- 다진 파와 마늘을 넣고 중불에 볶는다.
- 애호박을 넣고 살짝 익을 때까지 볶는다.
- 홍고추를 넣고 깨소금은 기호에 따라 넣는다.

3. 담아내기
- 접시에 호박나물을 담는다.

Ingredients

Main ingredients
Green Squash 400g (halfmoon, 0.5cm thick slices)

Sub-ingredients
Water 1/2C · Salt 1T · Cooking Oil 2T

Seasonings
Green Onion 1T (finely chopped) · Garlic 1t (minced) · Sesame Seeds 1/2T (ground)

Garnish
Red Chili 10g (bias-cut)

Steps

1. Salting green squash
- Make salt water.
- Add green squash and toss.
- After about ten minutes, drain off.

2. Stir-frying green squash
- Preheat pan and add cooking oil.
- Add green onion and garlic. Stir—fry over medium heat.
- Add green squash. Stir—fry until it is cooked slightly.
- Add red chili and ground sesame seeds to taste.

3. Plating
- Transfer to a plate.

귀띔 호박을 볶은 후 잠시 뚜껑을 덮어두면 호박에서 수분이 빠져나와 호박의 맛을 잘 살려준다.
Tips Covering the pan after stir—frying helps the squashes have more moisture and taste better.

죽순겨자채, 규아상, 땅콩호두장과

죽태(竹胎)라고도 하는 대의 땅속줄기에서 나오는 어린 싹을 새콤하게 무쳐
초록의 담쟁이 잎을 방석 삼아 앉힌 만두 한 접시 곁들여 먹고 대청마루에서
목침 베고 누워 쳐다보면 호사가 따로 없겠네…

Juksun-gyeoja-chae, Gyuasang, Ttangkong-hodu-janggwa

I nipped the baby sprouts from the bamboo stems underneath the ground and
seasoned them with sour sauce. Enjoying these sprouts with dumplings put on vivid
green ivy leaves would be just amazing. Yet, lying on the main floor and laying my
head on a wooden pillow, it would make me feel as if I was in heaven.

규아상

규아상은 해삼과 모양이 비슷하여 '미만두'라고 불린다. 궁중에서는 여름에 즐기던 음식으로
찐만두의 일종이다.

Gyuasang_ *Summer Dumpling with Cucumber and Beef*

Gyuasang is also called Mimandu because it looks like a sea cucumber. It is a type of steamed dumpling
which is enjoyed in the summer seasons in Korean Royal cuisine.

재료

주재료
오이 800g (5cm 길이로 자른 후 돌려깎아 채 썰기, 소금 1t) · 쇠고기 150g (간 것) · 건표고버섯 4장 (불려서 채 썰기) · 잣 1T

부재료
만두피 재료 · 밀가루 3C · 소금 1t · 물 9T · 식용유

양념
쇠고기와 버섯 양념 · 간장 2T · 참기름 2t · 설탕 1T · 다진 파 4t · 다진 마늘 2t · 통깨 2t · 후추 약간
초간장 · 간장 1T · 식초 1T · 물 1/2T · 설탕 1t · 잣가루 1/4t
고명 · 담쟁이 잎

만드는 법

1. 만두피 만들기
- 밀가루에 소금물을 넣고 반죽하여 비닐봉지에 담아 30분 정도 숙성시킨다.
- 긴 원통형으로 만든 후 만두피 한 개 분량으로 자른다.
- 자른 반죽을 밀대를 이용해 둥글게 민다.

2. 만두소 만들기
- 표고와 쇠고기는 양념장으로 양념 후 팬에 볶아준다.
- 채 썬 오이는 소금에 10분간 절였다가 물기를 짜고 기름 두른 팬에 볶아낸다.
- 오이, 쇠고기, 표고버섯을 한데 섞어서 소를 만든다.

3. 만두 빚기
- 만두피를 도마 혹은 손에 올려놓고 만두소 한 수저와 잣 하나를 넣는다.
- 만두피의 가장자리 끝에 물을 묻힌 후 반으로 접어 양쪽 끝을 제외하고 가장자리를 붙인다.
- 양쪽 끝을 삼각형 모양으로 접어 올려붙인다.
- 붙인 가장자리에 주름을 넣어 해삼 모양으로 빚는다.

4. 만두 찌기
- 찜기에 젖은 면보를 깔고 만두를 올려 10분간 쪄낸다.

5. 담아내기
- 접시에 만두를 담고 초간장은 작은 종지에 따로 담아낸다.

Ingredients

Main ingredients
Cucumber 800g (use of peel of skin, 5cm length, julienne, salt 1t) · Beef 150g (ground) · Dried Pyogo Mushroom 4ea (soaked, julienne) · Pine Nuts 1T

Sub-ingredients
Dumpling skin · Flour 3C · Salt 1t · Water 9T · Cooking Oil

Seasonings
For beef and mushroom · Soy Sauce 2T · Sesame Oil 2t · Sugar 1T · Green Onion 4t (finely chopped) · Garlic 2t (minced) · Sesame Seeds 2t · Black Pepper pinch
Dip · Soy Sauce 1T · Vinegar 1T · Water 1/2T · Sugar 1t · Pine Nuts 1/4t (ground)
Garnish · Ivy Leaves

Steps

1. Making dumpling skins
- Add salted water into flour and mix until dough is complete.
- Rest it wrapped in a plastic bag for about 30 min.
- Roll into sausage shape and slice into round balls.
- Roll them into flat circles.

2. Making dumpling fillings
- Marinate beef and mushroom.
- Stir-fry beef, mushroom and cucumber.
- Sprinkle salt over cucumber and set aside for 10 min.
- Squeeze out excess water.
- Lightly stir-fry cucumber with cooking oil.
- Combine fillings in a bowl, mix all together.

3. Shaping dumplings
- Place a dumpling skin on cooking board or hand.
- Put a spoonful of fillings and a pine nuts.
- Wet the edge of skin.
- Fold and seal edge but leave both ends unattached.
- Shape the ends into triangular shapes and seal them.
- Crimp the edge looking like sea cucumber.

4. Steaming
- Place a wet cloth and dumpling on a steamer.
- Steam for 10 min.

5. Plating
- Transfer to a plate.
- Serve with dipping sauce.

귀띔
Tips

1. 담쟁이 잎을 사용하여 찌면 만두가 서로 달라붙지 않아 좋다.
2. 만두피는 사서 써도 좋다.

1. When steaming dumplings, use ivy leaves. They help dumplings not stick.
2. You can buy dumpling skin in the market.

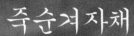

죽순겨자채

죽순과 편육, 오이, 배 등의 재료를 섞어 매콤한 겨자소스로 버무린 냉채이다. 톡 쏘면서
새콤달콤한 맛이 고기 음식과 잘 어울린다.

Juksun-gyeoja-chae

__ *Bamboo Shoots and Vegetables with Korean Mustard Dressing*

'Juksun-gyeoja-chae' is a cold dish with various vegetables and a spicy mustard dressing. The
refreshing-yet spicy-dressing also harmonizes well with meat dishes.

재료

주재료
통조림 죽순 50g • 양배추 2장 • 오이 1/2개
당근 1/4개 • 쇠고기 50g (양지머리) • 배 1/4개 (1×4×0.2cm
크기로 썰기) • 밤 3톨 • 달걀 1개

양념
연겨자 1½T • 식초 · 설탕 각 2T • 간장 1/2t
소금 약간 • 물 2T

고명
잣 약간

만드는 법

1. 재료 손질하기
- 죽순은 빗살무늬를 살려 썬다.
- 5분 정도 끓는 물에 삶아 석회질을 뺀다.
- 양배추와 오이, 당근, 배는 같은 크기로 썬다.
- 밤은 껍질을 벗기고 편으로 썬다.

2. 편육 준비하기
- 끓는 물에 쇠고기를 넣고 삶는다.
- 무거운 것으로 눌러 식힌다.
- 얇게 저며 다른 재료와 같은 크기로 썬다.

3. 지단 준비하기
- 달걀을 황 · 백으로 나누어 지단을 부친다.
- 다른 재료와 같은 크기로 썬다.

4. 버무려 내기
- 모든 재료를 겨자소스와 버무려 접시에 담고 잣을 올린
  다.

Ingredients

Main ingredients
Caned Bamboo Shoot 50g • Cabbage 2 leaves • Cucumber
1/2ea • Carrot 1/4ea • Beef 50g (brisket) • Korean Pear 1/4ea
(1×4×0.2cm sliced) • Chestnuts 3ea • Egg 1ea

Seasonings
Tubed Mustard 1½T • Vinegar 2T • Sugar 2T • Soy Sauce
1/2t • Salt pinch • Water 2T

Garnish
Pine Nuts pinch

Steps

1. Preparation
- Cut bamboo shoots maintaining the comb shape.
- Boil them in water for 5 minutes to remove lime.
- Cut cabbage, cucumber, carrot and pear into the same sizes.
- Peel chestnuts and slice them.

2. Cooking beef
- Cook beef in boiling water.
- Press the beef with heavy object while it gets cool down.
- Slice and cut beef into the same size as other ingredients.

3. Making Jidan
- Separate egg yolk and white and pan-fry.
- Cut them into the same sized pieces above.

4. Mixing all ingredients together
- Mix all ingredients together.
- Transfer to a plate and garnish with pine nuts.

귀띔 재료들은 차게 보관하였다가 버무려 낸다.
Tips Store cooked ingredients cold and then mix them right before serving.

땅콩호두장과

장과는 간장, 설탕, 물엿 등으로 양념하여 만든 조림음식으로 땅콩호두장과는 생땅콩과 호두를 조림장에 조려낸 반찬이다. 주로 밑반찬으로 쓰인다.

Ttangkong-hodu-janggwa

__ *Peanuts and Walnuts Simmered in Sweet Soy Sauce*

Janggwa means food boiled down in soy sauce and other seasonings such as sugar and corn syrup. Ttangkong-hodu-janggwa is a janggwa made of peanuts and walnuts, and it is usually served as a side dish.

재료

주재료
생땅콩 200g · 호두 100g

조림장
간장 3T · 물 3T · 설탕 2T · 물엿 3T
고추기름 1/2T · 식용유 1/2T · 참기름 1/2T
꿀 1/2T · 통깨 약간

고명
잣가루 약간

만드는 법

1. 생땅콩, 호두 손질하기
- 생땅콩은 끓는 물에 넣어 5분 정도 삶은 후 찬물에 씻어 물기를 제거한다.
- 호두는 끓는 물에 2분 정도 삶은 후 찬물에 씻어 물기를 제거한다.

2. 땅콩, 호두 볶기
- 두꺼운 팬에 식용유를 두르고 땅콩, 호두를 넣고 살짝 볶아준다.

3. 조림장에 조리기
- 분량의 양념장을 냄비에 넣고 끓으면 땅콩, 호두를 넣고 센 불에 조리다가 약불에서 서서히 저어가며 조린다.
- 거의 조려지면 꿀, 참기름, 통깨를 뿌린다.

4. 그릇에 담기
- 그릇에 담고 잣가루를 뿌린다.

Ingredients

Main ingredients
Fresh Peanuts 200g · Walnuts 100g

Seasonings
Soy Sauce 3T · Water 3T · Sugar 2T
Corn Syrup 3T · Red Chili Oil 1/2T
Cooking Oil 1/2T · Sesame Oil 1/2T
Honey 1/2T · Sesame Seeds pinch

Garnish
Pine Nuts pinch (ground)

Steps

1. Preparing fresh peanuts and walnuts
- Put fresh peanuts in boiling water for 5 min.
- Wash the peanuts with cold water and drain.
- Put walnuts in boiling water for 2 min.
- Wash the walnuts with cold water and drain.

2. Toasting peanuts and walnuts
- Slightly toasting peanuts and walnuts with cooking oil.

3. Simmering
- Mix all seasonings in a pot except sesame oil, honey and sesame seeds.
- When boil the seasonings add peanuts and walnuts.
- Bring it to boil and reduce the heat.
- Simmered the nuts until sauce gets lightly sticky.
- Add honey, sesame oil, sesame seeds and mix well.

4. Plating
- Transfer to a plate.
- Garnish with ground pine nuts.

비빔밥, 뭇국

비빔밥의 다른 이름인 골동반(骨同飯)에서 '골동'은 여러 가지
물건을 한데 섞는다는 뜻이다. 융합하는 능력이 경쟁력이 되는
시대에 동물성 식품인 음(陰)과 식물성 식품인 양(陽)의 반찬과
밥이 조화롭게 섞여 한 수저의 고추장으로 맛과 맛을 연결함이
한국인의 뛰어난 융합력을 보여주는 듯하다.

Bibimbap, Mu-guk

The 'Goldong', originating from the alternative name for Bibimbap,
'Goldongban', signifies a process of synthesizing diverse materials
that vary in nature. The combination of yin, yang side dishes —
deriving from animal and plant sources — and rice are perfected
with just a single spoon of Gochujang; it functions as a medium
that connects the different flavors of the ingredients. This synergy
of this food symbolizes the outstanding unity of Koreans, especially
when fusion takes a crucial role in the competitiveness of society,
in the modern world today.

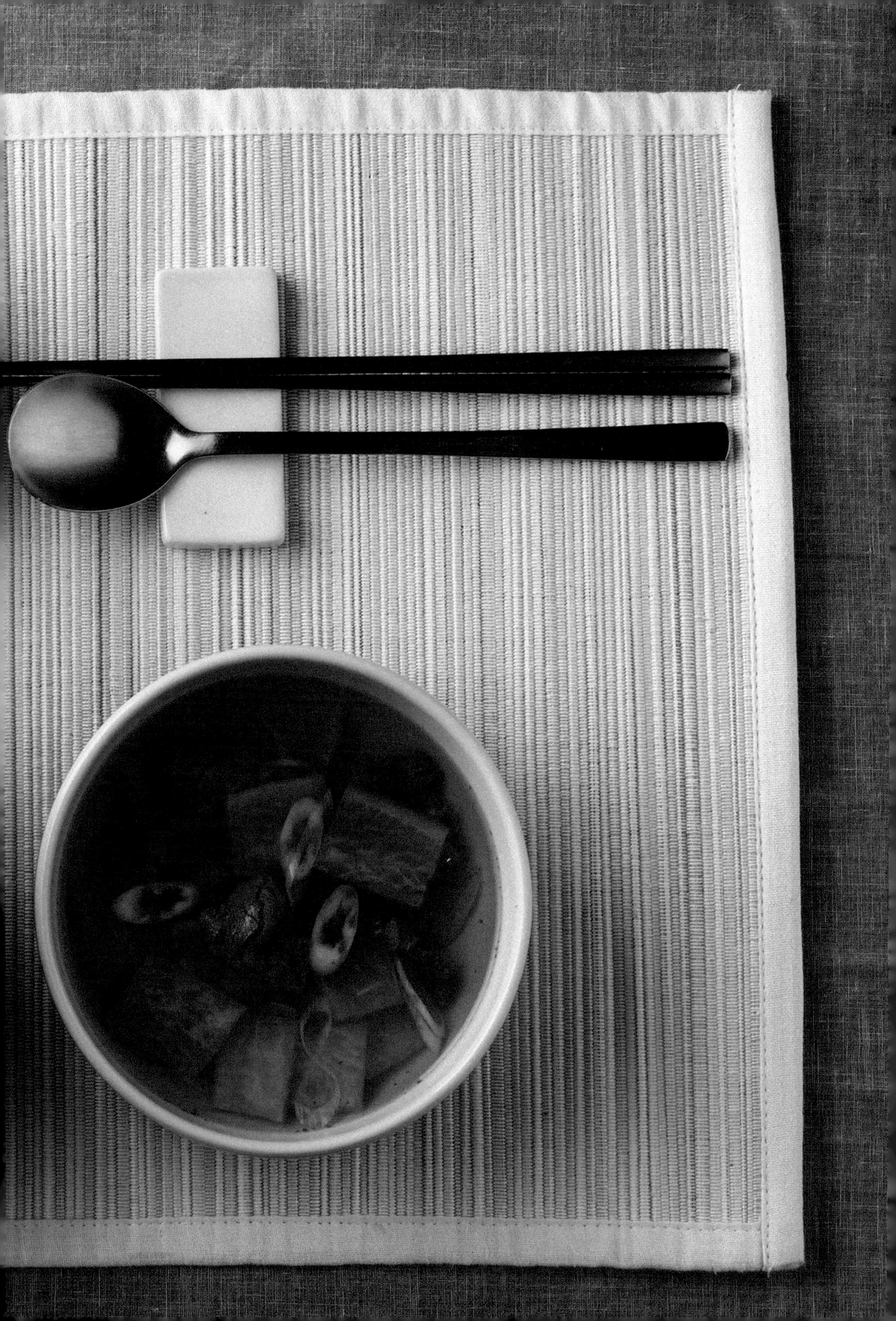

비빔밥

비빔밥에는 여러 가지 채소뿐만 아니라 다른 재료들도 많이 들어가 있어 그 자체만으로도 훌륭한 한 끼의 식사가 된다. 궁중에서는 골동반이라 불리었다. 섣달에 해가 바뀌기 전 남은 음식을 한데 모아 먹었던 데서 그 유래를 찾는다.

Bibimbap_ *Mix with Rice, Various Vegetables and Meat*

Since Bibimbap includes various kinds of vegetables and other ingredients, it can be nutritious meal in itself. In Korean Royal cuisine, Bibimbap is called 'Goldongban'-and it is said to have originated from the custom of mixing together all the left-overs before new year's.

재료

주재료
밥 4인분 · 도라지나물/고사리나물/시금치나물/콩나물 각 100g
(다듬기) · 애호박 100g (4등분, 얇게 썰기) · 쇠고기 80g (간 것)
건표고버섯 2개 (불려 채 썰기)
밥 밑양념 · 참기름 2T
콩나물 양념 · 참기름 2T · 소금 1/4t · 다진 파 1/4t
다진 마늘 1/4t
애호박 양념 · 소금 1/2t · 식용유 1/4t
고기 및 버섯 양념 · 간장 1T · 설탕 1/2t · 참기름 1/2t
다진 파 2t · 다진 마늘 1t · 후추 약간 · 깨소금 약간
고명(지단) · 달걀 1개 · 소금 약간
약고추장 재료 · 쇠고기 간 것 40g · 고추장 4T · 설탕 1/2t
식용유 1t · 참기름 1t · 꿀 1T
양념간장 재료 · 간장 2T · 물 1T · 설탕 1/2T · 깨소금 1/2t

만드는 법

1. 밥에 밑양념하기
- 밥과 참기름을 잘 섞는다.

2. 삼색나물
- 도라지나물, 고사리나물, 시금치나물 조리법 참조

3. 콩나물 무치기
- 냄비에 콩나물을 넣고 불에 익힌다. (물을 2~3T 넣는다.)
- 양념을 넣고 무친다.

4. 애호박 준비하기
- 애호박을 소금에 10분간 절여둔다.
- 면보로 물기를 제거한다.
- 기름을 두르고 살짝 볶아낸다.

5. 고기 및 버섯 준비하기
- 고기와 버섯을 5~10분간 양념에 재워둔다.
- 물을 조금 부어 저어가며 볶아낸다.

6. 약고추장 만들기
- 식용유를 두르고 쇠고기를 먼저 볶는다.
- 나머지 재료를 넣고 설탕이 녹을 때까지 젓는다.

7. 담아내기 및 고명 올리기
- 양념한 밥을 담고 그 위에 나물과 고기를 예쁘게 올린다.
- 지단 고명을 올리고 약고추장과 같이 낸다.

Ingredients

Main ingredients
Cooked Rice for 4 · Doraji-namul / Gosari-namul / Sigeumchi-
namul / Soybean Sprout 100g / each namul (trimmed) · Green
Squash 100g (quartered, sliced) · Beef 80g (minced) · Dried
Pyogo Mushroom 2ea (soaked, julienne)
Seasoning for rice · Sesame Oil 2T
Seasonings for soybean sprout · Sesame Oil 2T · Salt 1/4t
Green Onion 1/4t (finely chopped) · Garlic 1/4t (minced)
Seasonings for green squash · Salt 1/2t · Cooking Oil 1/4t
Seasonings for Beef & mushroom · Soy Sauce 1T · Sugar
1/2T · Sesame Oil 1/2t · Green Onion 2t (finely chopped) ·
Garlic 1t (minced) · Black Pepper pinch · Sesame Seeds pinch
(ground)
Garnish (Jidan) · Egg 1ea · Salt pinch
Yakgochujang · Beef 40g (minced) · Red Chili Paste
4T · Sugar 1/2t · Cooking Oil 1t · Sesame Oil 1t · Honey 1T
Seasoned soy sauce · Soy Sauce 2T · Water 1T · Sugar
1/2T · Sesame Seeds 1/2t (ground)

Steps

1. Seasoning rice
- Mix cooked rice and sesame oil.

2. Tri-color vegetable side dishes. (Samsaek-namul)
- Refer to Doraji-namul, Gosari-namul, Sigeumchi-namul
recipes.

3. Cooking soybean sprout
- Cook sprouts over heat in a pan. (Add 2~3T of water.)
- Mix well with seasonings.

4. Cooking green squash
- Salt green squash and set aside for 10 minutes.
- Remove moisture using cotton cloth.
- Stir-fry slightly with cooking oil.

5. Seasoning and stir-frying beef and mushroom
- Marinate for 5~10 min after seasoning.
- Stir-fry using a sauce pan slightly with water. (Keep stirring.)

6. Making Yakgochujang sauce
- Stir-fry beef with cooking oil first.
- Add remaining ingredients and stir until sugar is dissolved.

7. Plating and garnishing
- Place seasoned rice with vegetables and meat in a bowl.
- Garnish with Jidan on top and serve with sauce.

외국인도 빠져드는 한국밥상

귀띔
Tips
1. 약고추장 대신 양념간장을 쓰기도 한다.
2. 삼색나물 조리법은 84~89page를 참조한다.
3. 맑은국과 같이 낸다.
4. 영양 면에서도 손색없는 훌륭한 한 끼의 식사이다.

1. You can substitute Yakgochujang sauce with seasoned soy sauce.
2. Samsaek-namul recipes refer to pages 84~89.
3. Serve with a clear soup.
4. Nutritionally, it is a well-organized dish in itself.

고사리나물

고사리나물은 갓 캔 고사리나 말린 고사리를 데쳐서 양념하여 만드는 것이다. 일상반찬, 생일상, 명절 차례상에 주로 쓰이며 비빔밥의 재료로도 쓰인다.
고사리나물은 대표적인 어두운 색의 나물 반찬이다. 보관성을 좋게 하기 위해 수확한 후 건조하여 보관한다.

Gosari-namul_ *Stir-fried Fern Shoots*

Gosari-namul is blanched and seasoned fern shoots. It is usually served for everyday meals as well as meals for birthdays or Korean traditional commemorative rites for ancestors. It is also used as an ingredient Bibimbap (rice bowl mix with various vegetables and meat).
Gosari-namul (stir-fried fern shoots) is a representative vegetable dish on the Korean table. Koreans harvest and dry them to prolong storage.

재료

주재료
불린 고사리 100g (마른 나물 불리기 참조)

양념
식용유 1/2T · 간장 1T · 다진 파 1t
다진 마늘 1/2t · 참기름 1/2t · 깨소금 1/2t
물 2~3T · 후추 약간

만드는 법

1. 밑간하기
　-고사리에 양념을 넣고 주무른다.

2. 볶기
　-약불에서 볶는다.

3. 푹 익히기
　-약불에서 뚜껑을 덮고 물을 넣은 후 푹 익혀야 양념이
　　잘 밴다.

Ingredients

Main ingredients
Soaked Dried Fern Shoots 100g (refer to soaking dried vegetables in water)

Seasonings
Cooking Oil 1/2T · Soy Sauce 1T
Green Onion 1t (finely chopped)
Garlic 1/2t (minced) · Sesame Oil 1/2t
Sesame Seeds 1/2t (ground)
Water 2~3T · Black Pepper pinch

Steps

1. Pre-seasoning
　-Mix fern shoots and seasonings.

2. Stir-frying
　-Stir—fry over low heat.

3. Cooking thoroughly
　-After adding water, leave it on low heat with the lid on to cook thoroughly.

87

귀띔 고사리나물은 시금치, 도라지와 함께 삼색나물로 쓰인다.

Tips Gosari—namul (black) is a part of Samsaek—namul, tri—color vegetable side dish, along with Sigeumchi—namul (green) and Doraji—namul (white).

도라지나물

도라지는 사포닌 등의 영양소가 풍부하여 인삼에 버금가는 식재료로서, 도라지나물은 껍질 벗긴 통도라지를 가늘게 갈라서 양념하여 만든 것이다.

Doraji-namul_ *Stir-fried Balloon Flower Roots*

Doraji-namul is thin strips of balloon flower roots stir-fried and seasoned.
The balloon flower roots (Doraji) contains lots of saponin, which is widely known as Korean ginseng's healthy nutrient.

재료

주재료
도라지 100g (깐 것, 4등분) · 소금 1T (주무르기) · 물 5C (데치기)

양념
간장 1/4t · 다진 파 1/2t · 다진 마늘 1/4t · 식용유 1/2t · 물 3T (볶기) · 깨소금 1/2t · 참기름 1/2t · 소금 1/2t · 설탕 1/2t

만드는 법

1. 쓴맛 빼기
- 도라지에 소금을 넣고 바락바락 주물러 쓴맛을 뺀다.
- 물로 잘 헹군 후 건져낸다.
- 소금으로 주무른 도라지를 끓는 물에 데친 후 건져낸다.

2. 볶기
- 팬을 달구어 양념과 함께 볶아낸다.

Ingredients

Main ingredients
Balloon Flower Roots 100g (peeled, quartered) · Salt 1T (rubbing) · Water 5C (blanching)

Seasonings
Soy Sauce 1/4t
Green Onion 1/2t (finely chopped)
Garlic 1/4t (minced)
Cooking Oil 1/2t · Water 3T (for stir-frying)
Sesame Seeds1/2t (ground)
Sesame Oil 1/2t · Salt 1/2t · Sugar 1/2t

Steps

1. Rubbing
- Rub roots enough with salt to remove bitter taste.
- Rinse with water and drain.
- Blanch in boiling water and drain.

2. Stir-frying
- Stir-fry no. 1 in a pre-heated pan with seasonings.

89

귀띔 물기가 날아갈 정도로만 살짝 볶아야 도라지가 질겨지지 않는다.

Tips Stir-fry only until water evaporates.
Or it will turn out too tough.

시금치나물

시금치나물은 시금치를 빠르게 데쳐서 양념하여 무치는 나물이다. 참기름과 같은 지방성분을
보충함으로써 칼슘 섭취를 돕는다.

Sigeumchi-namul_ *Blanched and Seasoned Spinach*

Sigeumchi-namul is blanched and seasoned with spinach. Spinach should be blanched quickly to reduce
the loss of nutrients and vitamins. Adding sesame oil helps with the absorption of fat-soluble vitamins.

재료

주재료
시금치 150g (다듬기) · 물 5C (데치기) · 소금 1/2T (데치기)

양념
간장 1t · 다진 파 1/2t · 다진 마늘 1/4t
깨소금 1t · 참기름 1t · 식용유 1t · 소금 약간

만드는 법

1. 데치기
- 끓는 물에 소금을 약간 넣고 데쳐낸다.
- 차가운 물로 씻은 후 꼭 짠다.

2. 무치기
- 데친 시금치와 소금을 뺀 양념을 모두 넣고 잘 무쳐준다.
- 마지막으로 소금을 넣어 간을 맞춘다.

Ingredients

Main ingredients
Spinach 150g (trimmed) · Water 5C (blanching) · Salt 1/2T (blanching)

Seasonings
Soy Sauce 1t · Green Onion 1/2t (finely chopped) · Garlic 1/4t (minced) · Sesame Seeds 1t (ground) · Sesame Oil 1t · Cooking Oil 1t · Salt pinch

Steps

1. Blanching
- Add salt in boiling water and blanch spinach.
- Rinse with cold water and squeeze.

2. Seasoning
- Mix no. 1 and all seasonings well except salt.
- Add salt to season.

91

귀띔 시금치 데칠 때 소금을 넣으면 초록색을 유지할 수 있다.
Tips Addition of salt in boiling water helps maintain green color while blanching.

뭇국

뭇국은 맑은국을 대표하는 음식이다. 무의 시원한 맛이 국에 배어 고기와 잘 어울린다.
맑은국이기 때문에 매콤한 음식들과도 잘 어울린다.

Mu-guk_ *Radish Soup*

Radish gives a tasty flavor, so mu-guk goes well with heavy dishes. It also goes well with spicy dishes
since it is clear.

재료

주재료
쇠고기 200g (양지, 한입 크기) · 무 400g (한입 크기)

부재료
물 8C · 다시마 10g · 대파 60g (흰 부분, 어슷썰기)

고기양념
다진 마늘 1t · 참기름 1t · 후추 약간

국양념
국간장 1t · 소금 1t

만드는 법

1. 고기 밑양념하기
-양념을 모두 섞어 고기를 재워둔다.
2. 국 끓이기
-팬에 물 2T를 넣고 재워둔 고기를 볶는다.
-고기가 익으면 무를 넣고 잘 저어준다.
-물과 다시마를 넣는다.
-한소끔 끓으면, 다시마를 건져낸다.
-간장과 소금으로 간을 맞춘다.
-거의 다 끓었으면, 대파를 넣는다.
-다시 끓어오르면 불을 끈다.
3. 담아내기
-국그릇에 옮겨 담는다.

Main ingredients
Beef 200g (brisket, bite-sized) · Radish 400g (bite-sized)

Sub-ingredients
Water 8C · Kelp 10g · Green Onion 60g (white portion, bias-cut)

Seasonings for beef
Garlic 1t (minced) · Sesame Oil 1t
Black Pepper pinch

Seasonings for soup
Gukganjang 1t · Salt 1t

Steps

1. Seasoning beef
-Mix all seasonings and marinate beef for minutes.
2. Cooking soup
-Stir-fry marinated beef in a pan with 2T of water.
-When beef is cooked, add radish and stir well.
-Add water and kelp.
-Bring it to boil and take the kelp out.
-Season with soy sauce and salt.
-Lastly add green onion.
-Turn off heat, when starts boiling again.
3. Plating
-Transfer to a soup bowl.

김밥, 닭산적, 달걀북엇국

설레는 마음으로 지고 나선 소풍 길의 사이다와 김밥…
요즈음 어디서나 볼 수 있는, 다양한 재료와 모양으로 만든 한 끼 식사로 손색없는 김밥

Gimbap, Dak-sanjeok, Dalgyal-bugeo-guk

When I was young, I walked with joy when carrying bottles of sprits and rolls of Gimbap to the picnic. As times goes by, the various ingredients and shapes of Gimbap has changed Gimbap itself is a easily made and substantial meal.

닭산적

닭산적은 '닭꼬치'라고도 불린다. 만들기 간단하고, 영양소도 골고루 들어 있어 간식 혹은
술안주로 아주 좋다. 매콤한 맛은 입맛을 돋운다.

Dak-sanjeok_ *Spicy Chicken Skewer*

Dak-sanjeok is more popularly called Dak-kkochi in Korea. Dak means chicken and kkochi means
skewer in Korean. It is easy to make and nutritious. Dak-sanjeok is good for a snack and a side dish for
drinking. Its spicy taste brings out your appetite.

재료

주재료
닭고기 400g (안심/ 가슴살, 3cm 크기로 썰기)

부재료
가래떡 2개 (10cm 짜리 4등분) · 대파 2대 (3cm 길이로 썰기)

양념
닭 밑양념 · 청주 1T · 간장 1T · 생강즙 1t
후추 약간
가래떡 밑양념 · 참기름 1T · 간장 1T
고추장 양념장 · 고추장 · 물 · 청주 · 물엿 각 2T씩
간장 · 설탕 · 다진 마늘 각 1T씩 · 마른 고추 1개

고명
통깨 약간

만드는 법

1. 닭고기 밑양념하기
- 밑양념 재료를 섞어 15분 동안 닭을 재워둔다.
- 고추장 양념장을 만든다.
- 양념장의 1/2 분량을 닭에 넣고 섞는다.

2. 가래떡 데쳐 양념하기
- 가래떡을 끓는 물에 말랑하게 데쳐 밑양념한다.

3. 닭고기 익히기
- 팬에 기름을 두르고 양념한 닭고기를 볶는다.
- 다 익은 닭고기를 건져 한 김 뺀다.

4. 꼬치에 끼워 익히기
- 닭고기, 가래떡, 대파를 꼬치에 끼운다.
- 남은 양념장을 발라가며 팬에서 잠깐 구워낸다.

Main ingredients
Chicken 400g (Breast, 3cm cubes)

Sub-ingredients
Garae-tteok (stick rice cake) 2ea (10cm long, quater) · Green Onion 2ea (cut into 3cm)

Seasonings
For chicken · Rice wine 1T · Soy Sauce 1T · Ginger Juice 1t · Black Pepper pinch
For rice cake · Sesame Oil 1T · Soy Sauce 1T
For red chili paste sauce · Red Chili Paste, Water, Cheongju, Corn Syrup 2T / each · Soy Sauce, Sugar, Garlic (minced) 1T / each · Dried Red Chili 1ea

Garnish
Sesame Seeds pinch

Steps

1. Marinating chicken
- Mix the seasonings for chicken in a bowl. Leave it for 15 mins.
- Add chicken and set aside.
- Mix the seasonings for red chili paste.
- Use 1/2 the amount of red chili paste sauce with the chicken.

2. Blanching and seasoning rice cake
- Blanch rice cake in boiling water to slightly soften.
- Mix blanched garae-tteok with sesame oil and soy sauce.

3. Stir-frying chicken
- Add cooking oil in a pan and stir-fry the marinated chicken.
- Take the cooked chicken out and cool it down.

4. Skewering and cooking
- Skewer cooked chicken, garae-tteok, and green onion.
- Cook on a pan over medium-low heat brushing or pouring left sauce on it.

외국인도 빠져드는 한국밥상

97

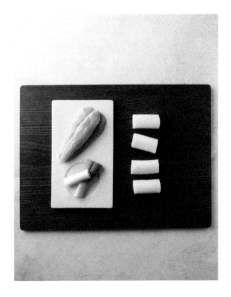

김밥

김밥을 만들 때 특정한 재료가 꼭 필요한 것은 아니므로 가능한 재료로 만든다. 일반적으로 한국 학생들은 소풍갈 때 점심으로 대부분 김밥을 싸간다. 또한 여행을 갈 때 간단한 식사로 김밥을 준비한다. 김밥은 한국인들이 가장 편하게 먹는 음식 중 하나이며 특히 학생들과 젊은 층이 좋아한다. 요즘은 다양한 모양과 재료를 넣어 만든 김밥으로 대중적인 사랑을 받고 있다.

Gimbap_ *Seaweed Roll with Rice and Various Fillings*

When making Gimbap, there aren't specific ingredients, so you can make it with any available ingredients. In general, when students go for a picnic in Korea, they bring Gimbap for lunch. Gimbap is popular as light lunch to bring for a short trip as well. It is also a favorite food for young Koreans favorite food. Gimbap can be made in various shapes and with different ingredients and is loved by most Koreans.

재료

주재료
쌀 360g (30분간 불리기) · 마른 김 4장 (살짝 굽기)

부재료
당근 50g (껍질 벗겨 채 썰기) · 단무지 70g (20cm 길이로 썰기, 0.7cm 두께) · 우엉 50g (채 썰기) · 시금치 80g (손질해서 소금물에 데치기) · 다진 쇠고기 80g · 달걀 2개 (풀기)

양념
밥 밑양념 · 소금 3/4t · 참기름 1T
당근 양념 · 소금 1/4t
우엉 양념 · 물 1/4C · 간장 1T · 설탕 1/2T · 다진 마늘 1/2t
쇠고기 밑양념 · 간장 2t · 다진 파 1t · 다진 마늘 1/2t · 통깨 1/2t · 참기름 1/2t · 후춧가루 약간
달걀 양념 · 소금 약간

만드는 법

1. 밥 짓고 양념하기
- 밥을 짓고 소금과 참기름으로 양념한다.

2. 쇠고기 양념하고 볶기
- 모든 양념을 섞어서 쇠고기에 재운다.
- 재운 쇠고기를 예열한 팬에서 볶는다.

3. 시금치 데치고 양념하기
- 시금치를 끓는 소금물에 데친다.
- 여분의 물기를 꼭 짜내고 소금으로 간한다.

4. 당근 볶기
- 채 썬 당근에 소금으로 간을 하고 예열한 팬에 기름을 두르고 볶는다.

5. 우엉 조리기
- 팬에 기름을 두르고 우엉을 볶는다.
- 물, 간장, 설탕, 마늘을 넣고 졸인다.

6. 지단 부치기
- 팬에 지단을 도톰하게 부친다.
- 폭 1cm 길이 20cm로 자른다.

7. 재료 넣고 싸기
- 발 위에 김 한 장을 올린다.
- 양념한 밥을 윗부분의 약 2cm 정도 남기고 김 위에 얇게 펼친다.
- 모든 재료(당근, 시금치, 단무지, 우엉, 지단, 쇠고기)를 밥의 가운데 놓고 발의 앞쪽 끝을 들어 올려 말아서 둥근 원형으로 만든다.

8. 담아내기
- 김밥을 10조각으로 썬다.
- 넓은 접시에 담아낸다.

Ingredients

Main ingredients
Rice 360g (soaked, 30 min.) · Dried Seaweed 4sheets (laver, lightly toasted)

Sub-ingredients
Carrot 50g (peeled, julienne) · Yellow Pickled Radish 70g (cut into 20cm long strips, 0.7cm thick) · Burdock Root 50g (julienne) · Spinach 80g (trimmed, blanched with salted water) · Beef 80g (ground) · Egg 2ea (separated, beaten)

Seasonings
For cooked rice · Salt 3/4t · Sesame Oil 1T
For carrot · Salt 1/4t
For burdock root · Water 1/4C · Soy Sauce 1T · Sugar 1/2T · Garlic 1/2t (minced)
For beef · Soy Sauce 2t · Green Onion 1t (finely chopped) · Garlic 1/2t (minced) · Sesame Seeds 1/2t · Sesame Oil 1/2t · Black Pepper pinch
For egg · Salt pinch

Steps

1. Cooking & seasoning rice
- Cook rice and season with salt and sesame oil.

2. Seasoning & stir-frying beef
- Mix all seasonings and marinate beef.
- Stir-fry marinated beef on a preheated pan.

3. Blanching & seasoning spinach
- Blanch spinach in salted boiling water.
- Drain and squeeze the excess water out.
- Season to taste with salt.

4. Stir-fry carrot
- Season with salt, stir-fry the carrot on a preheated pan with oil and set aside to cool down.

5. Simmering burdock root
- Stir-fry julienne burdock root with oil on a pan.
- Add water, soy sauce, sugar, garlic to simmer.

6. Making Jidan
- Separate white and yolk and season each to taste with salt.
- Pan-fry the beaten, salted egg in a layer.
- Cut them into 20cm long and 1cm thick strips.

7. Assembling & rolling up
- Place a sheet of laver on a bamboo mat or a cutting board.
- Spread seasoned rice in a thin layer on the laver leaving 2cm border on each side.
- Place all other ingredients (carrot, spinach, pickled radish, burdock root, fried egg strips and beef) on top of rice, lift up the edge of the bamboo mat and roll up the laver filled with ingredients.

8. Plating
- Cut the Gimbap into 10 pieces of flat rounds.
- Transfer them into the flat plate to serve.

외국인도 빠져드는 한국밥상

달걀북엇국

북어는 원기를 신속하게 회복시키고 알코올을 분해하며 간에 좋은 메티오닌, 타우린을 다량 함유하고 있다. 이 같은 이유로 한국 사람들은 음주 후 숙취를 해소하기 위해 북엇국을 먹는다.

Dalgyal-bugeo-guk_ *Egg & Dried Pollock Soup*

Pollock is rich in methionine and taurine, which help one's liver to swiftly refresh vitality and to dissolve alcoholic elements. Koreans enjoy it to relieve a hangover after drinking.

재료

주재료
북어채 60g

부재료
달걀 1개 • 대파 1개 (어슷썰기) • 홍고추 1개 (씨 빼기, 어슷썰기) • 물 6C

양념
참기름 1T • 소금 약간 • 후춧가루 약간

북어 밑양념
국간장 1T • 다진 마늘 1/2t • 후춧가루 약간

만드는 법

1. 북어 손질하여 양념하기
- 북어를 물에 적신다.
- 물기를 빼고 여분의 물기를 짠다.
- 양념과 잘 섞어둔다.

2. 달걀 풀기
- 달걀은 젓가락이나 거품기로 가볍게 풀어준다.

3. 양념북어 볶고 끓이기
- 냄비에 참기름을 넣고 북어를 넣어 볶는다.
- 물을 넣고 10분간 끓인다.

4. 기타 재료 넣기
- 하얀 국물이 우러나면 풀어놓은 달걀을 넣는다.
- 대파와 홍고추를 넣는다.
- 소금으로 간을 맞춘다.

5. 담아내기
- 국 대접에 옮겨 담는다.

Ingredients

Main ingredients
Torn dried Pollock 60g

Sub-ingredients
Egg 1ea (beaten) • Green Onion 1ea (bias-cut) • Red Chili 1ea (seeded, bias-cut) • Water 6C

Seasonings
For soup • Sesame Oil 1T • Salt pinch
Black Pepper pinch

For dried pollock
Gukganjang 1T • Garlic 1/2t (minced)
Black Pepper pinch

Steps

1. Seasoning dried pollock
- Dip into water to soften pollock.
- Drain and squeeze excess water out.
- Mix them well with seasonings and set aside.

2. Beating egg
- Beat egg with chopsticks or a whisk.

3. Stir-frying and boiling seasoned pollock
- Stir-fry seasoned pollock in a preheated pot.
- Pour water and bring it to boil for 10 minutes.

4. Adding other ingredients
- Once soup becomes milky-white color, add beaten egg.
- Add green onion and red chili.
- Season with salt to taste.
- Remove from the heat.

5. Plating
- Transfer to a soup bowl.

외국인도 빠져드는 한국밥상

101

잡채, 미역국, 배추겉절이

생일날 아침. 둘러앉은 밥상에는 하얀 쌀밥에 김 오르는
미역국, 잡채와 김치, 들기름 발라 구운 고소한 김. 태어나서
축복받는 존재의 이유가 새삼 느껴지는 날

Japchae, Miyeok-guk,
Baechu-geotjeori

Waking up on my birthday, I always found hot steaming Miyeok-
guk with a bowl of rice on the table. Japchae, Kimchi, and
laver toasted with perilla oil. Eating this dish with my family
surrounding me always reminded me of how blessed I was.

잡채

잡채는 손님접대용 음식으로 가장 널리 알려지고 오랫동안 사랑받고 있는 메뉴이다.
'잡'은 '여러 가지 종류', '채'는 '다양한 채소'라는 의미를 갖고 있다. 일반적으로 잡채는 최소한
5가지 색을 포함한다; 흰색(달걀 흰자, 양파; 서쪽), 황색(달걀 노른자; 중앙), 붉은색(당근; 남쪽),
녹색(오이; 동쪽), 검은색(고기, 표고버섯; 북쪽). 이러한 5가지 색은 오방색이라고도 불리며,
한국문화의 전통적인 관점에서 세계의 화합을 의미한다.

Japchae_ *Starch Noodles with Various Vegetables & Meat*

Japche is one of the well-known and long-loved dishes in the Korean party table setting. 'Jap' means
'many kinds of', and 'Chae' means 'various vegetables'. Generally japchae contains at least five colors;
white (egg white & onion; west), yellow (egg yolk; middle), red (carrot; south), green (cucumber; east)
and black (meat & Pyogo mushroom; north). These five colors, Obang Saek, stand for the harmony of
the world from the traditional perspective of Korean culture.

재료

주재료
당면 100g (불리기)

부재료
마른 표고버섯 2개 (불리기, 길이로 채 썰기, 0.5cm 두께) · 쇠고기 150g (채 썰기, 0.5cm 두께) · 당근 1/4개 (길이로 채 썰기) · 양파 1/4개 (길이로 채 썰기) · 오이 1/2개 (껍질만 채 썰기, 소금에 절이기 (물 1/2C, 소금 1t, 물기 제거하기) · 달걀 1개 (흰자, 노른자 분리하기)

양념
당면 양념 · 간장 1½T · 설탕 1½T · 참기름 1T
쇠고기 · 버섯 양념 · 간장 1T · 설탕 1/2T · 다진 대파 2T · 다진 마늘 1T · 참기름 1T · 깨소금 1½T
후추 약간
당근 양념 · 다진 대파 1t · 다진 마늘 1/2t · 참기름 1/2t · 소금 1/2t

만드는 법

1. 당면 불리기
- 당면은 30분간 미지근한 물에 넣어 불린다.

2. 쇠고기와 버섯 양념에 재우기
- 쇠고기 · 버섯 양념을 각각 1/2씩 쇠고기와 버섯에 넣고 버무린다.

3. 당근 데치기
- 끓는 소금물에 당근을 살짝 데친다.
- 물기 제거한 후 당근 양념을 넣고 버무린다.

4. 재료 볶기
- 예열한 팬에 식용유를 넣는다.
- 센 불에 양파, 오이, 당근을 각각 살짝 볶는다.
- 쇠고기와 버섯을 볶는다.

5. 달걀지단 만들기
- 달걀 노른자에 약간의 소금을 넣고 섞는다.
- 예열한 팬에 식용유를 두르고 지단을 얇게 부친다.
- 0.5cm 두께로 채 썬다.
- 흰자도 노른자와 같은 방법으로 한다.

6. 당면 데치기와 양념하기
- 불린 당면을 끓는 물에 투명하게 데쳐 식힌다.
- 식힌 당면에 당면 양념을 넣고 버무린다.

7. 버무리기와 간하기
- 모든 재료를 골고루 섞어준다.
- 설탕, 참기름, 간장은 기호에 맞게 조절한다.

8. 담아내기
- 따뜻할 때 접시에 담아낸다.

Ingredients

Main ingredients
Starch Noodle 100g (soaked)

Sub-ingredients
Dried Pyogo Mushroom 2ea (soaked, julienne, 0.5cm thick) · Beef 150g (julienne, 0.5cm thick) · Carrot 1/4ea (julienne lengthwise) · Onion 1/4ea (julienne lengthwise, rinse) · Cucumber 1/2ea (peel only, julienne, soak in 1/2C of water with 1t of salt & drain) · Egg 1ea (separate into yolk and white)

Seasonings
For starch noodle · Soy Sauce 1½T · Sugar 1½T · Sesame Oil 1T
For beef & mushroom · Soy Sauce 1T · Sugar 1/2T · Green Onion 2T (finely chopped) · Garlic 1T (minced) · Sesame Oil 1T · Sesame Seeds 1½t (ground) · Black Pepper pinch
For carrot · Green Onion 1t (finely chopped) · Garlic 1/2t (minced) · Sesame Oil 1/2t · Salt 1/2t

Steps

1. Soaking starch noodle
- Soak starch noodle for 30 minutes in warm water.

2. Marinating beef & mushroom
- Mix all seasonings for beef & mushroom.
- Use 1/2 of the seasoning for beef.
- Use 1/2 of the seasoning for mushroom.

3. Blanching carrot
- Blanch the carrot in boiling salted water.
- Drain water and season.

4. Stir-frying ingredients
- Preheat a pan and add cooking oil.
- Stir-fry onion, cucumber, carrot separately on high heat for seconds.
- Transfer to a plate to cool it down.
- Repeat steps for mushroom and beef.

5. Making jidan
- Add some salt to egg yolk and stir well.
- Thinly pan-fry on preheated oil-coated pan.
- Julienne the jidan as 0.5cm thick.
- Repeat steps for egg white.

6. Blanching & seasoning starch noodle
- Blanch the starch noodle until it becomes clear and cool it down.
- Add starch noodle marinade sauce and mix well.

7. Mixing & seasoning
- Mix well all ingredients.
- Add sugar, sesame oil and soy sauce to taste.

8. Plating
- Transfer to a plate. Serve warm.

배추겉절이

배추겉절이는 묵은 김치가 떨어지거나 신김치에 싫증 날 때 해 먹는 즉석김치로 배추속대를 살짝
절였다가 길게 쭉쭉 찢어서 무치면 배추의 싱싱한 맛이 살아 있어 신선하고 개운하다. 김치와
샐러드의 중간 형태라 할 수 있다.

Baechu-geotjeori_ *Quick Method Kimchi*

Baechu-geatjeori can be easily prepared when you're tired of eating sour and old Kimchi. Just leave
the cabbage in salted water for a while, mix with the sauce, and you can enjoy the refreshing taste of it
right away. It is a mix between regular Kimchi and salad.

재료

주재료
어린 배추 1/2통 (길이로 썰기, 5cm 폭) · 물 5C · 굵은소금
1/4C

부재료
쪽파 50g (잘게 썰기, 5cm 길이) · 대파 100g (어슷썰기) · 홍
고추 1개 (고추씨 빼서 어슷썰기)

양념
겉절이 양념 · 고춧가루 3/4C · 새우젓 2½T (곱게 다지
기) · 설탕 2T · 간장 1T · 생강즙 3/4T · 다진 마늘 2½T ·
참기름 1T · 통깨 1½T

고명
통깨 약간

만드는 법

1. 소금에 절이기
- 물에 소금을 넣고 고루 섞어준다.
- 손질한 배추는 소금물에 넣고 한 시간 동안 둔다.
- 절여진 배추는 가볍게 물로 헹궈 물기를 뺀다.

2. 양념 만들기
- 겉절이 양념을 섞는다.

3. 버무리기
- 절인 배추에 겉절이 양념을 넣고 골고루 섞는다.
- 부재료를 넣고 골고루 섞는다.
- 참기름과 통깨를 넣고 버무린다.

4. 담아내기
- 접시에 담고 통깨를 뿌린다.

Ingredients

Main ingredients
Young Korean Cabbage 1/2ea (lengthwise-cut, 5cm width) · Water
5C · Salt 1/4C

Sub-ingredients
Scallion 50g (chop, 5cm) · Green Onion 100g (bias-cut) · Red Chili
1ea (halved, seeded, bias-cut)

Seasonings
For geotjeori · Red Chili powder 3/4C · Salted Shrimp 2½T
(finely chopped) · Sugar 2T · Soy Sauce 1T · Ginger 3/4T
(minced) · Garlic 2½T (minced) · Sesame Oil 1T · Sesame Seeds
1½T

Garnish
Sesame Seeds pinch

Steps

1. Salting young Korean cabbage
- Dissolve salt in water.
- Soak the trimmed cabbage in the brine for an hour.
- Rinse the cabbage slightly and drain off.

2. Making seasoning
- Mix all geotjeori seasonings well.

3. Mixing
- Add geotjeori seasoning into salted cabbage.
- Add sub-ingredients. Mix well.
- Add sesame oil and sesame seeds. Mix well.

4. Plating
- Transfer to a plate and sprinkle sesame seeds.

귀띔
Tips

1. 배추를 절이는 동안 적어도 두 번은 뒤집어줘야 고루 잘 절여진다.
2. 배추겉절이는 불고기, 된장찌개, 밥과 어울리는 음식이다.
   이러한 상차림은 영양학적으로나 맛으로나 완벽한 차림이 된다.

1. Toss young cabbage upside down at least two times during salting to aid better texture.
2. Baechu-geotjeori goes well with Bulgogi, Doenjang-jjigae and cooked rice.
   These make a perfect table in nutrition and taste.

생일상차림

한국 사람들은 생일을 맞는 가족을 축하하기 위해 흰쌀밥에 미역국,
고기와 전, 잡채 등으로 생일상을 정성껏 차려 식사를 같이한다.
그중 미역국은 부기를 가라앉히고 혈압을 조절해 주므로 아이를 낳은
산모가 산후조리를 위해 먹는 음식이며, 한국인에 있어 미역국은
태어난 날을 상징한다.

Birthday Table

Koreans prepare hearty dishes to celebrate the birthday of a family
member. All the family members gather and each enjoys a bowl of white
rice with Miyeok-guk, meat. Jeon, and Japchae. Because this dish can
subsiding swelling and regulate blood pressure. Miyeok-guk (seaweed
soup in Korean) is well known for recovering health among mothers who
just gave birth to a child. Miyeok-guk is also the symbol of a birthday in
Korean culture.

미역국

한국의 엄마들은 출산 후 미역국을 먹는다. 미역은 피를 맑게 해주고 수유에 도움을 주는 음식으로 잘 알려져 있기 때문이다. 한식에서는 미역뿐만 아니라 김, 다시마, 파래 등과 같은 다양한 종류의 해조류를 사용하고 있다.

Miyeok-guk_ *Seaweed Soup*

Korean mothers usually have miyeok-guk after giving birth, because seaweed is well-known for its effect on blood purification and breast-feeding.

Korean dishes have various kinds of seaweed dishes used not only for miyeok (dried seaweed), but also for gim (laver), dasima (kelp), parae (sea lettuce), etc.

재료

주재료
마른미역 10g (찬물 5컵에 불리기, 거품나지 않을 때까지 씻기, 한 입 크기로 자르기)

부재료
쇠고기 100g (양지머리, 잘게 썰기) ・ 물 5~7C

양념
소고기용 밑양념 ・ 다진 마늘 1/2t ・ 참기름 1t ・ 소금 1/2t ・ 국간장 1/2T
국간장 2T ・ 다진 마늘 1/2t

만드는 법

1. 쇠고기와 미역 볶기/국물내기
- 냄비를 예열한다.
- 쇠고기에 참기름, 마늘, 국간장을 넣고 고르게 볶는다.
- 미역을 넣고 볶는다.

2. 끓이기와 양념하기
- 물을 붓고 중간불로 20분 정도 끓인다.
- 기호에 맞춰 국간장을 넣어 간한다.
- 한소끔 끓인다.

3. 담아내기
- 뜨거울 때 국 대접에 담아낸다.

Ingredients

Main ingredients
Dried Seaweed 10g (soak, water 5C, wash until it has no bubbles, bite-size cut)

Sub-ingredients
Beef 100g (brisket, small bite-size cut) ・ Water 5~7C

Seasonings
For beef ・ Garlic 1/2t (minced) ・ Sesame Oil 1t ・ Salt 1/2t ・ Gukganjang 1/2t
For soup ・ Gukganjang 2t ・ Garlic 1/2t (minced)

Steps

1. Stir-frying beef and seaweed
- Preheat a pot.
- Marinade beef and stir—fry well.
- Add seaweed and stir—fry throughly on medium heat.

2. Boiling & seasoning
- Add water and bring it to boil. Reduce the heat to medium for 20 minutes.
- Add Gukganjang and garlic to taste.
- Bring it to boil.

3. Plating
- Transfer to a soup bowl. Serve hot.

귀띔
Tips
1. 불린 미역을 참기름으로 충분히 볶아야 풍미가 풍성해지고 깊은 맛을 내며 부드러운 미역국을 먹을 수 있다.
2. 국간장으로 간한 미역국이 소금으로 간한 것보다 더 깊은 맛과 색을 내지만, 소금을 사용해도 무방하다.

1. Stir—fry seaweed with sesame oil throughly to get a more nutty, deep and silky taste.
2. Gukganjang (soy sauce for soup) is much better than salt for creating deep flavor and color. However, you may use salt instead, as well.

삼계탕, 양배추김치, 마늘장아찌

무더운 여름에 삼계탕을 먹어 더위를 이긴 조상들의 풍습으로부터
내려온 지혜로운 밥상

Samgye-tang, Yangbaechu-kimchi, Maneul-jangajji

The table with Samgye-tang is a table of wisdom. This dish has been passed down from our ancestors who stood the heat of the summer with this special dish.

삼계탕

삼계탕은 어린 닭의 배 속에 찹쌀, 마늘, 대추, 인삼을 넣고 물을 부어 오래 끓인 음식으로
계삼탕이라고도 한다. 여름철 보신음식으로 꼽는다. 원래는 영계를 백숙으로 푹 곤 것을
'영계백숙'이라 했는데 여기에 인삼을 넣어 계삼탕이라 하다가 지금은 삼계탕으로 굳어졌다.

Samgye-tang_ *Chicken Ginseng Soup with Rice Stuffing*

Samgye-tang is a summer favourite for Koreans. This is especially enjoyed on Sam-bok (which is the
hottestdays in summer). During this hottest period of a year, it is easy to lose one's appetite and stamina
due to perspiration. This well-harmonized dish of ginseng, chicken and other various ingredients help
revive those losses. It might sound unusual for foreigners, however, Koreans believe that they can beat
the heat by eating very hot food.

재료

주재료
영계 1마리 (500~600g) · 물 8C

부재료
불린 찹쌀 1/2C · 수삼 1뿌리 · 대추 5개 · 밤 3개 (껍질 벗긴 것) · 은행 4개 (껍질 벗긴 것) · 마늘 6개

양념
대파 2T (송송 썰기) · 소금 약간 · 후추 약간

만드는 법

1. 밑 준비
- 닭은 찬물에 담가 핏물을 뺀 후 깨끗이 씻어둔다.
- 찹쌀은 씻은 후 물에 담가 불린다.
- 수삼은 깨끗이 씻어 칼로 뇌두(윗부분)를 잘라낸다.
- 대추와 껍질 깐 밤은 씻어서 건져낸다.
- 은행은 기름 두른 팬에 볶아 껍질을 벗겨낸다.
- 대파는 송송 썰어둔다.

2. 채우기
- 닭 배 속에 찹쌀, 수삼, 대추, 밤, 은행, 마늘을 채워 넣고 다리를 오므린다.

3. 끓이기
- 냄비에 닭을 넣고 닭이 잠길 정도로 물을 붓고 끓인다.

4. 담아내기
- 삼계탕은 예열한 그릇에 담아낸다.
- 소금, 후추, 송송 썬 대파를 같이 낸다.

Ingredients

Main ingredients
Young Chicken 1ea (500~600g) · Water 8C

Sub-ingredients
Sweet rice (Glutinous rice) 1/2C (wash, soak at least for 2 hours) · Ginseng root 1ea · Jujubes 5ea · Chestnuts 3ea (peeled) · Gingko nuts 4ea (peeled) · Garlic 6cloves

Seasonings
Green onion 2T (chopped) · Salt pinch · Black Pepper pinch

Steps

1. Preparing chicken
- Clean chicken with water and drain.

2. Stuffing
- Stuff chicken with sweet rice, ginseng root, chestnuts, jujubes, gingko nuts and garlic cloves.
- Cross chicken legs to keep stuffings in while being boiled.

3. Boiling
- Add no. 2 in a pot with 8 cups of water.
- Bring it to boil.
- Lower the heat and let simmer until chicken is cooked thoroughly.

4. Plating
- Transfer to a preheated soup bowl.
- Serve with salt, black pepper and chopped green onion.

외국인도 빠져드는 한국밥상

115

귀띔
Tips

1. 끓어오를 때 기름기를 걷어낸다.
2. 깍두기와 함께 내면 좋다.

1. While being boiled, skim off floating fat and impurity.
2. It goes well with kkakdugi (radish Kimchi).

양배추김치

봄과 여름 사이에 햇배추가 재배되기 전 배추 대신 손쉽게 구할 수 있는 양배추로 담근 김치이다.
아삭하고 상큼한 맛의 양배추와 오이가 잘 어우러져 훌륭한 식감을 즐길 수 있는 별미 김치이다.

Yangbaechu-kimchi_ *Cabbage Kimchi*

Yangbaechu Kimchi is made of cabbage rather than Napa cabbage. Kimchi is generally made of Napa cabbage, but regular cabbage is also a good substitute for Napa cabbage especially when Napa cabbage is out of season between spring and summer. In addition, cabbage and cucumber are well-matched ingredients; together they provide a refreshing flavor as well as crunchy texture.

재료

주재료
양배추 2kg (4×4cm 썰기) · 오이 500g (3cm 길이, 길이로 3등분, 씨 제거)

절임
소금 3T · 물 3C

부재료
실파 100g (3cm 길이 썰기)

양념
마늘 60g · 양파 100g · 생강 20g · 액젓 1/2C · 설탕 1T · 고춧가루 1/2C · 소금 약간

만드는 법

1. 절이기
- 오이, 양배추 썬 것에 소금물을 부어 30분간 절인다.
- 절여진 채소는 물에 한번 헹구어 물기를 뺀다.

2. 양념 만들기
- 믹서에 양념을 넣고 곱게 간다.
- 양념에 고춧가루 1/2C을 섞는다.

3. 버무리기
- 물기 뺀 양배추와 오이에 양념을 넣고 잘 버무린다.
- 실파를 넣고 살짝 버무린다.
- 부족한 간은 소금으로 한다.

4. 담아내기
- 접시에 담아낸다.

Ingredients

Main ingredients
Cabbage 2kg (4×4cm slice) · Cucumber 500g (3cm slice, cut into 3 pieces length, remove the core)

Brine
Salt 3T · Water 3C

Sub-ingredients
Small Green Onion 100g (cut in 3cm length)

Seasonings
Garlic 60g · Onion 100g · Ginger 20g · Fish Sauce 1/2C · Sugar 1T · Red Chili Powder 1/2C · Salt pinch

Steps

1. Salting cabbbage and cucumber
- Dissolve salt in water.
- Soak the prepared cabbage and cucumber in the brine for 30 minutes.
- Rince the salted cabbage and cucumber once, and drain off all the remaining water.

2. Seasoning
- Mix and grind all the ingredients from seasoning list except red chili powder.
- After grinding all the mixture, add the 1/2 cup of red chili powder to the mixture.

3. Mixing
- Mix prepared cabbage and cucumber with the seasoning mixture.
- Add the small green onion.
- Adjust the seasoning with salt.

4. Serving
- Transfer to a plate and serve it.

117

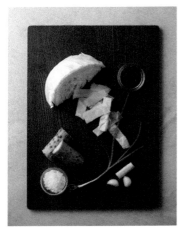

귀띔
Tips

1. 양배추김치에 사과 등 과일을 갈아 넣으면 맛과 영양이 좋아진다.
2. 실온에서 하루 정도 숙성시키면 더욱 맛있다.

1. Korean pear and apple juice can be added to seasoning mixture. It will increase flavor and nutrition.
2. If Cabbage Kimchi stays in room temperature overnight, it will get fermented and taste better.

깻잎찜

깻잎은 향긋한 향이 매력적인 채소로서 생으로 먹기도 하고 절임이나 찜으로 먹기도 한다.

Kkaennip-jjim_ *Steamed Perilla Leaves with Seasoning*

Perilla leaves have a savory scent and can be pickled and steamed.

재료

주재료
깻잎 2묶음 (24장) (깨끗이 씻어 물기 빼기)

양념
대파 1/2T (길이 2cm 정도로 채 썰기) • 마늘채 1t • 간장
1½T • 물 1½T • 깨소금 1/2T • 참기름 1t • 고춧가루 1t

만드는 법

1. 양념 만들기
　- 분량의 양념 재료를 섞어 양념장을 만든다.
2. 찌기
　- 턱이 약간 올라오는 접시에 깻잎과 양념을 켜켜이 놓는
　　다.
　- 김이 오른 찜통에 접시를 놓고 약 5분간 찐다.

Ingredients

Main ingredients
Perilla Leaves 2bundles ((24leaves) wash and drain thoroughly)

Seasonings
Green Onion 1/2T (2cm length, julienne)
Garlic 1t (julienne) • Soy Sauce 1½T
Water 1½T • Sesame Seeds 1/2T (ground)
Sesame Oil 1t • Red Chili Powder 1t

Steps

1. Making seasoning
　- Mix all seasoning in a bowl.
2. Steaming
　- Lay perilla leaves over a plate rubbing seasoning layer by layer.
　- Steam for about 5 minutes.

119

귀띔
Tips

1. 마른 보로 싼 뚜껑을 덮으면 물이 떨어지는 것을 방지해 준다.
2. 깻잎을 오래 찌면 질겨진다.

1. Covering the steam pot lid with a dry cotton cloth prevents the steam from leaking.
2. Over-steaming makes it tougher.

마늘장아찌

마늘은 한국의 식생활에서 빼놓을 수 없는 양념이다. 마늘은 비위를 따뜻하게 하고 기를 돋우며 살충의 효능이 있다고 한다. 고기를 먹을 때 마늘을 함께 먹는 것은 마늘이 고기의 누린내를 없애줄 뿐만 아니라 맛 또한 좋게 하며 육류에 들어 있는 비타민 B₁의 흡수를 돕기 때문이다.

Maneul-jangajji_ *Pickled Garlic*

Garlic is an important traditional ingredient in Korean cuisine. According to traditional oriental medicine documents, garlic warms one's stomach, energizes one's spirit, and enhances men's virility. Garlics are often served with meat because they remove meat's bad smell and boost meat's taste. Nutritionally, garlics help with the absorption of vitamin B_1.

재료

주재료
깐 마늘 1.5kg

부재료
물 1½C · 간장 6T · 식초 2½C
설탕 1/2C · 소금 4T

만드는 법

1. 물에 소금과 설탕을 넣고 잘 저어준다.
2. 1에 분량의 간장과 식초를 넣는다.
3. 통에 마늘을 넣고 2를 붓고 뚜껑을 닫는다.
4. 일주일간 둔다.
5. 일주일 후 국물만 따라낸다.
6. 끓인 후 식힌다.
7. 식힌 국물을 다시 마늘이 들어 있는 통에 붓는다.
8. 다시 2주일 정도 둔다.
9. 3~5를 반복한다.
10. 국물과 함께 낸다.

Ingredients

Main ingredients
Garlic 1.5kg (peeled)

Sub-ingredients
Water 1½C · Soy Sauce 6T · Vinegar 2½C · Sugar 1/2C · Salt 4T

Steps

1. Dissolve salt and sugar in water.
2. Add soy sauce and vinegar.
3. Put garlic in a jar and pour no. 2 in.
4. Set aside for 1 week.
5. Drain pickling water only.
6. Boil the pickling water and cool it down.
7. Pour the pickling water back in to the jar.
8. Set aside for 2 weeks again.
9. Repeat steps 3 to 5 in once again.
10. Serve it with sauce.

제육구이, 조개탕

가격이 저렴하면서도 고기 맛과 포만감을 동시에
줄 수 있어 주머니가 가벼운 사람들의 심리적
만족감까지 더해주었던 제육구이와 조개탕
매운맛을 시원하게 조개탕이 잡아주니, 음식
사이에도 보이지 않는 궁합이 있다.

Jeyuk-gui, Jogae-tang

Jeyuk-gui and Jogae-tang are two dishes that have not
only been affordable, but also give satisfaction of taste
and fullness to those with light wallets.
The refreshing and zesty taste of Jogae-tang keeps the
balance of spicy Jeyuk-gui, and implies there is an
hidden complementary relationship among foods.

삼겹살과 쌈 상

삼겹살은 한국인이 가장 사랑하는 음식 중 하나이다. 삼겹살뿐 아니라 다른 고기구이 음식들은 쌈채소와 각종 쌈장류와 잘 어울린다. 쌈은 복을 싸먹는다는 의미를 지니기도 한다.

Samgyeopsal & Ssam-sang_ *Grilled Pork Belly and Korean Barbecue Table Setting*

Samgyeopsal is one of the most popular foods in Korea. It is often wrapped in assorted vegetables as wraps and different sauces. Wrapping is called-Ssam in Korean and ssam means carrying luck as well.

재료

주재료
삼겹살 200g

부재료
쌈채소
상추 · 깻잎 · 양배추 데친 것 · 고추 · 마늘편

쌈장
된장 4T · 다진 마늘 1/2t · 통깨 1t · 참기름 1½t · 고추장
1t · 다진 파 1t
모든 재료를 잘 섞어낸다.
약고추장 (비빔밥 조리법 참조)
쇠고기 간 것 40g · 설탕 1/2t · 참기름 1t · 식용유 1t · 고
추장 4T

만드는 법

1. 굽기
  -삼겹살을 팬에 굽거나 숯불에 굽는다.
2. 쌈장 만들기
  -분량의 재료를 골고루 섞는다.
3. 약고추장 만들기
  -비빔밥 조리법 참조

Ingredients

Main ingredients
Pork Belly 200g (per portion)

Sub-ingredients
Ssamchaeso (assorted vegetables for wrapping)
Lettuce · Perilla Leaves · Blanched Cabbage · Green
Chili · Garlic (slice)

Dip
Ssamjang
Soybean Paste 4T · Garlic 1/2t (minced) · Sesame Seeds
1t · Sesame Oil 1½t · Red Chlil Paste 1t · Green Onion 1t (finely
chopped)
Yakgochujang (refer to Bibimbap recipe)
Beef 40g (minced) · Sugar 1/2t · Sesame Oil 1t · Cooking Oil
1t · Red Chili Paste 4T

Steps

1. Pan-frying or grilling pork belly
  -Pan-fry or grill to serve.
2. Making ssamjang
  -Mix all ingredients well.
3. Making yakgochujang
  -Refer to bibimbap recipe.

귀띔 삼겹살구이는 따로 소금이나 새우젓, 참기름을 곁들여 먹을 수 있다.
Tips Samgyeopsal-gui can be served with salt, sesame oil or salted shrimp.

제육구이

제육구이는 대표적인 고추장구이 음식이다. 고추장이 돼지 누린내를 잡아주며, 고기의 맛과
식감이 살아 있는 음식이다.

Jeyuk-gui_ *Grilled or Stir-fried Pork with Red Chili Paste Sauce*

Jeyuk-gui is a cornerstone of red chilli paste dish. Red chilli paste removes the smell from pork meat
and the texture and flavor of the meat becomes very good and tasty.

재료

주재료
돼지고기 500g (삼겹살, 한입 크기)

부재료
양파 100g (길이로 썰기) • 대파 40g (어슷썰기) • 식용유 1T

양념
고추장 2¼T • 굵은 고춧가루 1½T • 간장 1T • 설탕 1½T • 참기름 1T • 다진 마늘 ¾T • 청주 2¼T • 후추 약간 • 물 2¼T

고명 Garnish
통깨 1t 혹은 채 썬 파

만드는 법

1. 양념 만들기
- 큰 그릇에 양념을 모두 넣고 잘 섞는다.

2. 고기 재우기
- 돼지고기를 양념과 고루 섞고 양파, 대파를 넣어준다.
- 15분 정도 재워둔다.

3. 볶아내기 / 굽기
- 기름을 약간 두르고 팬을 달궈 고기를 볶아낸다.
- 혹은 숯불이나 그릴을 사용하여 잘 굽는다.

4. 담아내기 및 고명 올리기
- 그릇에 옮겨 담고, 통깨를 약간 뿌려준다.

Ingredients

Main ingredients
Pork 500g (belly, sliced, bite-sized)

Sub-ingredients
Onion 100g (julienne) • Green Onion 40g (bias-cut) • Cooking Oil 1T

Seasonings
Red Chili Paste 2¼T • Red Chili Powder 1½T (coarse) • Soy Sauce 1T • Sugar 1½T • Sesame Oil 1T • Garlic ¾T (minced) • Rice Wine 2¼T • Black Pepper pinch • Water 2¼T

Garnish
Sesame Seeds 1t or Green Onion, julienne

Steps

1. Making sauce
- Mix all seasonings in a big bowl.

2. Marinating
- Add pork into no. 1 and mix well. Add vegetables.
- Marinate for 15 minutes.

3. Stir-frying/ grilling
- Stir-fry the marinated meet mixture on a preheated pan with cooking oil.
- Or cook thoroughly using grill or broiler.

4. Plating and garnishing
- Transfer to a plate with a pinch of sesame seeds on the top.

127

귀띔
Tips

제육구이는 쌈과 같이 내면 좋다. 상추, 깻잎 등의 쌈채소와 맑은국, 밥을 같이 내면 아주 푸짐한 한상차림이 된다.

Jeyuk-gui goes well with Ssam (assorted vegetables). It becomes a generous serving with assorted vegetables, such as lettuce and perilla leaves, clear soup and rice.

조개탕

조개탕은 대표적인 해산물 맑은 탕이다. 끓이는 시간이 짧을 뿐 아니라 준비도 간단하다. 시원한
국물 맛은 풍부하게 함유된 타우린과 핵산에서 나오는데, 피로회복과 피부에도 좋다.

Jogae-tang_ *Clear Clam Soup*

Jogae-tang is a cornerstone of clear soup with seafood. This clear clam soup is quick and easy to
prepare. Clams not only give refreshing taste but also are good for tiredness relief.

재료

주재료
조개 400g (깨끗이 씻기) · 물 6C

부재료
마늘 1톨 (다지기 혹은 저미기) · 쪽파 2~3대 (3cm 길이로 썰기) · 홍고추 1/2개 (어슷썰기)

양념
소금 약간

만드는 법

1. 조개 국물내기
- 조개와 물을 한데 넣고 끓인다.
- 조개의 입이 벌어지면 불을 끈다.
- 조개와 국물을 분리해 놓는다.

2. 국물과 부재료 끓이기
- 조개 국물을 면보에 거른다.
- 한소끔 끓인다.
- 마늘과 소금을 넣는다.
- 쪽파와 고추를 넣고 불을 끈다.

3. 담아내기
- 국그릇에 익힌 조개를 담고 다시 끓여낸 육수를 붓는다.

Ingredients

Main ingredients
Clam 400g (washed) · Water 6C

Sub-ingredients
Garlic 1clove (minced or sliced) · Scallion 2~3pc (3cm length cut) · Red Chili 1/2ea (bias-cut)

Seasonings
Salt to taste pinch

Steps

1. Making clam stock
- Add clams and water in a pot and bring to boil.
- When clams open, turn off heat.
- Separate clam and stock.

2. Boiling stock with sub-ingredients
- Filter stock using cotton cloth.
- Bring the stock to boil.
- Add garlic and season with salt.
- Add scallion and red chili. Turn off the heat after few seconds.

3. Plating
- Put cooked clams in a soup bowl and pour the soup over.

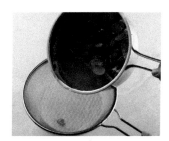

귀띔
Tips

1. 조개를 소금물에 넣어두어 해감시킨다. (1시간 정도)
2. 조개를 너무 오래 끓이면 질겨진다.
3. 뚜껑을 덮고 끓이면 비린내가 난다.

1. Leave the clam in the salted water so as to clean the sediment. (for about an hour)
2. Do not boil clams too long, in oder to prevent them from toughening.
3. Do not cover while cooking, in oder to evaporate the off—seafood—odor.

갈비찜 *Galbi-jjim* / 녹두전 *Nokdu-jeon* / 시금치된장국 *Sigeumchi-doenjang-guk* / 닭감자찜 *Dak-gamja-jjim* / 해물완자전 *Haemul-wanja-jeon* / 숙주나물 *Sukju-namul* / 오이나물 *Oi-namul* / 온면(잔치국수) *On-myeon (Janchi-guksu)* / 해물파전 *Haemul-pajeon* / 양지편육 *Yangji-pyeonyuk* / 타락죽 *Tarak-juk* / 닭잣죽무침 *Dak-jatjuk-muchim* / 콩자반 *Kong-jaban* / 나박김치 *Nabak-kimchi* / 김치찌개 *Kimchi-jjigae* / 달걀찜 *Dalgyal-jjim* / 감자소고기조림 *Gamja-sogogi-jorim* / 두부조림 *Dubu-jorim* / 떡국 *Tteok-guk* / 양배추오이생채 *Yangbaechu-oi-saengchae* / 북어구이 *Bugeo-gui* / 동치미 *Dongchimi* / 양지곰탕 *Yangji-gomtang* / 대하찜 *Daeha-jjim* / 깍두기 *Kkakdugi* / 더덕구이 *Deodeok-gui* / 김치 *Kimchi* / 보쌈 *Bossam* / 보쌈무생채 *Bossam-mu-saengchae* / 우거지된장국 *Ugeoji-doenjang-guk* / 김치만둣국 *Kimchi-mandu-guk* / 고기버섯산적 *Gogi-beoseot-sanjeok* / 장김치 *Jang-kimchi*

가을
/
겨울

갈비찜, 시금치된장국,
녹두전

예전엔 '반살미'라 하여 갓 혼인한 신랑이나
신부를 일갓집에서 처음으로 초대하여
대접하곤 했었다. 갈비찜과 전 몇 가지,
그 집안을 대표할 수 있는 밑반찬 올리고
사촌, 당숙 인사시키던 지난 시간이 새삼
그리워진다.

Galbi-jjim, Sigeumchi-doenjang-guk, Nokdu-jeon

There is a tradition called "Bansalmi" in which a newlywed groom invites the bride to his house for the first time and treats her with a hearty meal. This meal consists of Galbi-jjim, several Jeons, and the number of side dishes representing his house. I miss the time when the bride came to visit the house to say hello to her husband's cousins.

갈비찜

갈비찜은 소나 돼지의 갈비를 갖은 양념하여 만든 찜류의 음식이다.
갈비는 지방이 많은 조직이므로 조리 시 지방을 적절히 제거해 주는 것이 좋다.

Galbi-jjim_ *Rib Stew with Sweet Soy Sauce*

Galbi-jjim is beef or pork ribs braised with various ingredients in sweet soy sauce.
Since ribs may have much fat on it, it is advised to check and remove the fat before cooking.

재료

주재료
쇠갈비 1kg (5cm 토막) · 물 5C → 육수 3C

부재료
무 150g (3cm 깍둑썰기) · 양파 100g (분리되지 않게 3cm 결대로 썰기) · 당근 100g (3cm 깍둑썰기) · 마른 표고버섯 5개 (물에 불려 기둥 제거) · 밤 5톨 (껍질 깐 것) · 은행 5톨 (껍질 깐 것)

양념
간장 5T · 배즙 6T · 설탕 2T · 청주 1T · 다진 마늘 1T · 다진 파 2T · 생강즙 1/2t · 후춧가루 약간 · 참기름 1T · 깨소금 1T

만드는 법

1. 갈비 삶고 육수내기
- 갈비를 찬물에 1시간 이상 담가 핏물을 제거한다.
- 물을 넣고 30분간 끓인다.
- 꺼내어 기름덩어리는 제거하고, 2cm 간격으로 칼집을 넣는다.
- 육수의 기름은 걷어낸다.

2. 조리기
- 양념을 모두 섞어 양념장을 만든다.
- 삶은 갈비와 양념장 ⅔를 넣고 잘 섞어준다.
- 3컵의 육수를 붓는다.
- 약 15분간 중불에서 조린다.

3. 채소 익히기
- 은행을 제외한 부재료를 모두 넣고 약불에서 더 조린다.

4. 마무리하기
- 거의 다 되었을 때 은행을 넣는다.
- 불을 세게 하여 잠시 두었다가 불을 끈다.
- 찜기에 담아낸다.

Ingredients

Main ingredients
Beef Ribs 1kg (5cm cubes) · Water 5C → Stock 3C

Sub-ingredients
Radish 150g (3cm cubes) · Onion 100g (3cm lengthwise, head intact) · Carrot 100g (3cm cubes) · Dried Pyogo Mushroom 5ea (soaked in water, stem removed) · Chestnuts 5ea (skinned) · Gingko Nuts 5ea (skinned)

Seasonings
Soy Sauce 5T · Pear Juice 6T · Sugar 2T · Rice Wine 1T · Garlic, minced 1T ·
Green Onion 2T (finely chopped) · Ginger Juice 1/2t · Black Pepper, ground pinch · Sesame Oil 1T · Sesame Seeds 1T (ground)

Steps

1. Boiling ribs and making stock
- Soak ribs in cold water for more than one hour to remove blood.
- Boil ribs with water for 30 min.
- Remove fat chunks on ribs & give slits on 2cm interval.
- Skim fat off in stock.

2. Braising
- Make sauce by adding all the seasonings.
- Add boiled ribs and ⅔ of sauce and stir well.
- Pour 3cups of stock in.
- Braise for about 15 min. on a medium heat.

3. Cooking vegetables
- Add all the sub-ingredients in except gingko nuts and simmer on a low heat.

4. Adding gingko nuts and plating
- When almost cooked, add gingko nuts.
- Increase heat for few seconds and turn off.
- Transfer to a stew bowl to serve.

135

귀띔
Tips
1. 무와 당근의 모난 부분을 부드럽게 깎아주면 모서리가 부서져 국물이 지저분해지는 것을 방지할 수 있다.
2. 갈비찜은 보온이 가능한 그릇에 담아 국물을 자작하게 부어 따뜻하게 담아내는 것이 좋다.

1. Tear or trim off edges of radish and carrot to make a clearer stew.
2. Serve in a thermo plate or stew bowl with enough amount of sauce to enjoy Galbi-jjim warm.

녹두전

녹두를 갈아 채소나 고기와 함께 부치는 전이다.
빈대떡으로도 불리는 녹두전은 특히 막걸리와 잘 어울리는 음식 중 하나이다.

Nokdu-jeon_ *Mung Bean Dough Pan-fry*

Nokdu-jeon, which is also called "Bindaetteok", is one of Koreans' favorite "jeon". It is often served with Makgeolli, Korean traditional rice wine. "Jeon" is a unique cooking method in Korea. Nokdu-jeon is made from mung bean flour, vegetables, and meat; ingredients are mixed into mung bean dough and it is fried on a griddle.

재료

주재료
녹두 1C (불리면 2½C) · 물 2/3C (녹두 갈 때)

부재료
김치 100g (잘게 썰기) · 숙주 50g (4~5cm 길이로 자르기) · 다진 돼지고기 100g · 홍고추 1/2개 (어슷썰기) · 멥쌀가루 1/2C · 식용유

돼지고기 양념
소금 1/2t · 다진 마늘 1/2t · 생강즙 1/2t · 참기름 1t · 후추 약간

만드는 법

1. 녹두 거피하기
- 녹두를 물에 5시간 이상 담가둔다.
- 껍질을 벗기고 깨끗이 씻는다. 물기를 뺀다.

2. 녹두 갈기
- 1번의 녹두와 물을 섞은 뒤 블렌더를 이용하여 간다.

3. 돼지고기 볶기
- 돼지고기에 모든 양념을 넣고 밑간을 한다.
- 보슬보슬하게 볶는다.

4. 반죽 만들기
- 볼에 녹두 간 것, 볶은 돼지고기, 김치, 숙주, 홍고추, 멥쌀가루를 섞어 반죽을 만든다.

5. 부치기
- 팬에 기름을 넉넉히 두른다.
- 예열 후, 노릇노릇해질 때까지 지진다.

Ingredients

Main ingredients
Mung Bean 1C (2½C when soaked) · Water 2/3C (for grinding mung bean)

Sub-ingredients
Kimchi 100g (chopped) · Mung Bean Sprout 50g (cut into 4~5cm pieces) · Pork 100g (minced) · Red Chili 1/2ea (bias-cut) · Rice flour 1/2C · Cooking Oil

Seasonings for pork
Salt 1/2t · Garlic 1/2t (minced) · Ginger Juice 1/2t · Sesame Oil 1t · Black Pepper pinch

Steps

1. Hulling mung beans
- Soak mung beans in water for more than 5 hours.
- Hull and wash off. Drain well.

2. Grinding mung beans
- Combine no. 1 and water. Use blender to grind.

3. Stir-frying pork
- Mix all seasonings with minced pork.
- Stir-fry until moisture evaporates.

4. Making mixture
- Mix ground mung bean, fried pork, mung bean sprout, Kimchi, red chili, and rice flour into a bowl.

5. Pan-frying
- Coat a pan with enough amount of cooking oil.
- Pre-heat and pan-fry the mixture until golden brown.

137

시금치된장국

된장을 풀어 시금치를 넣고 끓인 국이다.
가정에서 즐겨 먹는 된장국 중 하나이다.

Sigeumchi-doenjang-guk

_ *Spinach Soup with Soy Bean Paste*

Sigeumchi-doenjang-guk is a spinach soup with soy bean paste.
It is a very popular Korean dish and many Korean families enjoy it at home.

재료

주재료
시금치 200g · 된장 3T

멸치육수 재료
마른 멸치 20g · 물 8C

양념
고추장 2t · 다진 마늘 1T · 대파 1대 · 소금 약간

만드는 법

1. 멸치육수 내기
- 물기 없이 팬에 멸치를 볶는다.
- 8컵의 물을 붓고 10분간 끓인다.

2. 시금치 데치기
- 끓는 물에 소금을 약간 넣고 시금치를 데쳐낸다.

3. 국 끓이기
- 체를 이용해 된장, 고추장을 육수에 풀어 토장국을 만든다.
- 충분히 끓으면 시금치와 마늘을 넣는다.
- 한소끔 끓인다.
- 마지막으로 대파를 넣고, 소금으로 간을 한다.

Ingredients

Main ingredients
Spinach 200g · Soy Bean Paste 3T

Ingredients for anchovy broth
Dried Anchovy 20g · Water 8C

Seasonings
Red Chili Paste 2t · Garlic 1T (minced) · Green Onion 1ea · Salt pinch

Steps

1. Making anchovy broth
- Toast anchovy in a pan without moisture.
- Add 8 cups of water and boil for 10 minutes.

2. Blanching spinach
- Add a little salt in boiling water and blanch spinach.

3. Boiling soup
- Add soy bean paste and red chili paste in anchovy broth using a seive.
- When it is boiled enough, add blanched spinach and garlic.
- Bring it to boil.
- Add green onion and season with salt.

귀띔 시금치를 데칠 때는 수산을 날려 보내기 위해서 반드시 뚜껑을 연다.
Tips Blanch spinach with the lid off in order to evaporate its acidic substances.

닭감자찜, 해물완자전,
숙주나물, 오이나물

특별한 날이 아니어도 모두 둘러앉아 먹는 한상차림
식구라서 좋은 날

Dak-gamja-jjim, Haemul-wanja-jeon, Sukju-namul, Oi-namul

Even a regular day could turn into a special one when we all ate together,
just as we always did.

닭감자찜

닭고기에 감자를 넣어 매운 양념으로 만든 음식이다.

Dak-gamja-jjim_ *Spicy Chicken Stew with Potato*

Dak-gamja-jjim is a spicy chicken stew with potato.

재료

주재료
닭 500g (5cm 크기로 자르기) · 소금 1T · 후춧가루 1t · 식용
유 2T

부재료
양파 1개 (4등분, 삼각모양썰기) · 감자 300g (3cm 크기로 썰어
모서리 다듬기) · 당근 100g (3cm 크기로 썰어 모서리 다듬기) ·
청 · 홍고추 각 2개 (어슷썰어 헹구기) · 대파 1대 (어슷썰기) ·
물 3C
*향신채소(선택사항)

양념
굵은 고춧가루 3T · 다진 마늘 2T · 다진 파 3T · 간장
1T · 고추장 2~3T · 생강즙 1t · 설탕 1T · 물엿 1T · 청주
1T · 소금 1t

고명
통깨 약간

만드는 법

1. 유장 처리하기
- 닭을 깨끗이 씻어 소금, 후추를 뿌려 밑간한다.
- 식용유 1T를 넣고 고루 주물러둔다.
2. 감자, 당근 준비하기
- 감자, 당근을 소금물에 2분간 삶아 찬물에 헹궈준다.
3. 닭 굽기/채소 익히기
- 팬을 달궈 기름 1T을 두르고 감자, 당근, 양파를 살짝 볶
는다.
- 선택 1: 250℃로 10분간 예열한 오븐에 유장 처리한 닭
을 넣고 온도를 150~170℃로 줄여 20분가량 익힌다.
- 선택 2: 달궈진 팬에 향신채소*와 기름을 넣고 볶다가,
닭을 넣고 굴려가며 익혀준다. 익힌 닭을 체에 담고 끓
는 물을 부어 기름기를 제거한다.
*향신채소: 양파 1/4쪽, 대파 15g, 마늘 5쪽, 생강 5g, 건고
추 2개
4. 양념장 만들기
- 갖은 양념을 고루 섞는다.
5. 찜하기
- 닭, 채소, 양념장 1/2, 물을 넣고 센 불에서 약 30분간 끓
인다.
- 국물이 반으로 줄면 나머지 양념을 넣는다.
- 불을 줄여 뚜껑을 닫고 푹 익힌다.
- 국물이 1/3 정도로 졸아들면 청 · 홍고추, 대파를 넣고
익힌다.
6. 그릇에 담기 및 고명 올리기
- 그릇에 담고 고명으로 통깨를 뿌려낸다.

Ingredients

Main ingredients
Chicken 500g (cut in 5cm) · Salt 1T · Black Pepper 1t
(ground) · Cooking Oil 2T

Sub-ingredients
Onion 1ea (quartered, cut into triangles)
Potato 300g (cut in 3cm, tourned)
Carrot 100g (cut in 3cm, tourned)
Red, Green Chili 2ea/each (bias-cut, rinsed)
Green Onion 1ea (bias-cut) · Water 3C
Aromatic Vegetables Optional

Seasonings
Red Chili Powder 3T (coarse) · Garlic 2T (minced)
Green Onion 3T (minced) · Soy Sauce 1T
Red Chili Paste 2~3T · Ginger Juice 1t · Sugar 1T · Corn Syrup
1T · Rice Wine 1T · Salt 1t

Garnish
Sesame Seeds pinch

Steps

1. Marinating chicken
- Wash chicken and season with salt and black pepper.
- Rub seasoned chicken adding 1T of cooking oil.
2. Blanching potato & carrot
- Blanch potato & carrot in boiling water with some salt for 2 mins.
and rinse.
3. Roasting chicken and pre-cooking vegetables
- Preheat a pan with 1T of cooking oil and slightly stir-fry potato,
carrot and onion.
- Option 1: Preheat oven to 250℃ for 10 minutes and roast chicken
for 20minutes reducing heat down to 150~170℃.
- Option 2: Preheat a pan with little cooking oil and stir-fry aromatic
vegetables*. Add chicken to cook. Strain excess fat by pouring boiling
water over.
*Aromatic Vegetables: Onion 1/4ea, Green Onion 15g, Garlic 5
cloves, Ginger 5g, Dried Red Chili 2ea
4. Mix all seasonings to make sauce.
5. Simmering
- Boil chicken, vegetables, water, and a half of sauce over high heat
for 30 minutes.
- When reduced by half, add the remaining sauce.
- Reduce the heat and cover lid to simmer.
- When reduced by 1/3, add chilis and green onion.
- Turn off heat after few seconds.
6. Plating and garnishing
- Transfer to a stew bowl and sprinkle sesame seeds.

귀띔
Tips

1. 채소를 비슷한 크기로 썰어야 잘 익고 모양새도 좋다.
2. 닭고기를 오븐에 구웠다가 소스에 조리면 기름이 빠지면서 껍질이 쫄깃쫄깃해져 씹는 맛이 좋아진다.

1. Cut vegetables into same-sized pieces to reduce cooking time and for better appearance.
2. Roasting chicken prior to simmering reduces fat and gives chewier texture.

해물완자전

새우나 오징어와 같은 해산물을 주재료로 반죽하여 둥글게 빚어 부친 전이다.

Haemul-wanja-jeon

__ *Pan-fried Seafood and Vegetable Pancake*

Haemul-wanja-jeon is a pan fried round-shaped mixture with seafoods such as shrimp and squid.

재료

주재료
새우살 50g (잘게 다지기) · 오징어 100g (잘게 다지기)

부재료
양파 1/2개 (잘게 다지기) · 소금 1/2T · 청 · 홍 · 황피망 1/4
개씩 (잘게 다지기) · 달걀 1개 · 밀가루 4T · 식용유 2T

초간장
간장 2T · 물 1T · 식초 1/2T · 설탕 1t

만드는 법

1. **양파 준비하기**
   - 양파는 소금을 넣고 절인 뒤 체에 밭친다.
   - 마른 면보로 물기를 꼭 짠다.
2. **재료 고루 섞기**
   - 주재료와 부재료를 볼에 담아 잘 섞는다.
3. **전 부치기**
   - 달궈진 팬에 반죽을 한 숟가락씩 떠 놓아 부친다.
   - 동그랗게 모양을 잡아가며 노릇하게 지진다.
4. **그릇에 담기**
   - 접시에 옮겨 담고 초간장을 곁들여낸다.

Ingredients

Main ingredients
Shrimp 50g (finely chopped) · Squid 100g (finely chopped)

Sub-ingredients
Onion 1/2ea (finely chopped) · Salt 1/2T · Red, Green, Yellow Bell
Pepper 1/4ea/each (finely chopped) · Egg 1ea · Flour 4T · Cooking
Oil 2T

Dip
Soy Sauce 2T · Water 1T · Vinegar 1/2T · Sugar 1t

Steps

1. **Preparing onion**
   - Sprinkle salt over onion and set aside for 5 minutes.
   - Squeeze out excess water with cotton cloth.
2. **Mixing**
   - Add all ingredients except cooking oil and mix well.
3. **Pan-frying**
   - Pre-heat a pan with cooking oil.
   - Pan-fry nice and crispy the mixture by a spoonful.
   - Shape them using a spoon.
4. **Plating**
   - Transfer to a plate to serve. Serve it with dipping sauce.

귀띔 반죽이 너무 되면 전이 딱딱하므로 물을 약간 섞는다.
Tips If the mixture gets too thick add small portion of water before frying.

숙주나물

숙주는 많은 아시아 국가에서 사용되는 식재료이다. 중국이나 태국 등에서는 숙주를 볶거나 생것으로 따뜻한 국물과 먹는다. 하지만 한국에서는 숙주를 살짝 데쳐 소금과 참기름으로 간하여 나물로 먹는다.

Sukju-namul_ *Blanched and Seasoned Mung Bean Sprouts*

Mung bean sprout is a well-known vegetable in Asia. Most Korean people cook blanched sukju (mung bean sprout) for namul with salt and sesame oil. However, in China or Thailand, they use sprouts for stir-frying or serve fresh sprouts with hot soup.

재료

주재료
숙주 200g

양념
소금 1t · 다진 파 1t · 다진 마늘 1/2t · 참기름 1/2t · 깨소
금 1t · 국간장 1/2t

고명
통깨 약간

만드는 법

1. 데치기
- 숙주를 끓는 물에 소금을 넣고 살짝 데친다.
- 건져낸 숙주의 물기를 꼭 짠다.

2. 양념 만들기
- 갖은 양념을 고루 섞는다.

3. 그릇에 담기 및 고명 올리기
- 그릇에 담아낸다.
- 통깨를 약간 뿌린다.

Ingredients

Main ingredients
Mung Bean Sprout 200g

Seasonings
Salt 1t · Green Onion 1t (finely chopped) · Garlic
1/2t (minced) · Sesame Oil 1/2t · Sesame Seeds 1t
(ground) · Gukganjang 1/2t

Garnish
Sesame Seeds pinch

Steps

1. Blanching
- Blanch mung bean sprouts in boiling salted water.
- Squeeze out excess water by hands.

2. Seasoning
- Mix blanched mung bean sprouts and all seasonings well.

3. Plating
- Transfer to a plate to serve.
- Garnish with sesame seeds.

귀띔
Tips

1. 숙주나물은 식초를 넣어 무치기도 하며 데친 미나리를 섞어서 무치기도 한다.
2. 콩나물과 함께 대표적인 흰색 나물반찬이다.
   제사상에도 쓰인다.

1. To be an appetizer, add 1t of vinegar, replacing the sesame oil and blanched Korean watercress.
2. Sukju-namul is one of the representative white-color namul dishes in Korea. This is served for memorial services, as well.

오이나물

오이는 여름채소로서 풍부한 과즙과 상큼한 맛으로 널리 사랑받는 식재료이다.
오이나물은 얇게 썬 오이를 소금에 살짝 절여 볶아낸 것이다.

Oi-namul_ *Stir-fried Cucumber*

Oi means a cucumber, which is a popular ingredient especially in summer because it is juicy and has a refreshing taste. Oi-namul is sliced cucumbers lightly pickled in salt and stir-fried.

재료

주재료
오이 400g (0.2~0.3cm 두께로 썰기)

부재료
굵은소금 1T (오이 절임용) • 물 1C (오이 절임용)

양념
소금 약간 • 다진 파 2t • 다진 마늘 1t • 참기름 1t • 깨소금 1t

고명
실고추 약간

만드는 법

1. 오이 손질
- 굵은소금으로 문질러 씻는다.
- 썬 오이를 소금물에 10분 정도 절인다.
- 절인 오이는 면보에 꼭 짜 물기를 제거한다.

2. 볶기
- 달군 팬에 기름을 두르고 오이, 다진 파, 다진 마늘을 넣고 센 불에서 재빨리 볶는다. (취향에 따라 소금을 추가할 수 있다.)
- 불을 끄고 깨소금, 참기름을 넣어 골고루 섞어준다.

3. 식히기
- 볶은 오이를 쟁반에 넓게 펴서 빨리 식힌다.

4. 담아내기
- 그릇에 담고 고명으로 실고추를 올린다.

Ingredients

Main ingredients
Cucumber 400g (0.2~0.3cm thick slices)

Sub-ingredients
Coarse Salt 1T (soaking) • Water 1C (soaking)

Seasonings
Salt pinch • Green Onion 2t (finely chopped) • Garlic 1t (minced) • Sesame Oil 1t • Sesame Seeds 1t (ground)

Garnish
Silgochu pinch

Steps

1. Preparing cucumber
- Rub cucumber with coarse salt.
- Slice cucumber and soak them in brine for 10 min.
- Squeeze excess water out of cucumber using cotton cloth.

2. Stir-frying
- In a pre-heated pan, put little cooking oil and add prepared cucumber, green onion and garlic. Stir-fry them quickly over high heat. (may season more.)
- Turn off heat and add sesame oil and sesame seeds.

3. Chilling
- Spread the stir-fried cucumber on a tray and let them cool down.

4. Plating and garnish
- Serve with julienned dried red chili.

귀띔 / Tips

1. 오이의 돌기 부분을 굵은소금으로 비벼 씻으면 불순물을 제거할 뿐 아니라 색도 더 선명해진다.
2. 오이를 센 불에서 재빨리 볶아 펼쳐서 식혀야 더 아삭하고 색이 선명하다.

1. Rubbing cucumber with salt helps remove tiny thorns and enhance green color.
2. Stir-frying over high heat maintains its crunchy texture.

양지편육, 온면(잔치국수), 해물파전

"국수 언제 줄래?" 하며 결혼식을 비유해 말하는 잔치국수
통과의례에 빠지지 않았던 음식

Yangji-pyeonyuk, On-myeon (Janchi-guksu), Haemul-pajeon

People often ask "When are you inviting me to have Guksu?", This is a soup noodle that is figuratively compared to marriage. It is also a dish that represents a ritual of passage.

온면 (잔치국수)

온면은 잔치국수라고도 불리며, 잔치국수는 결혼식이나 회갑 등의 큰 잔치에 오신 많은 손님들을
대접함으로써 붙여진 이름이다.
'국수'는 무병장수를 기원하는 음식으로 우리 조상들의 풍류적인 기질을 엿볼 수 있다.

On-myeon (Janchi-guksu)_ *Noodles with Clear Broth*

On-myeon is also called Janchi-guksu (Janchi mean feast or banquet) because this noodle dish has been
eaten for special occasions like weddings ceremony or 60th birthdays (Ilwan gab) or other celebrations
in Korea. The meaning of noodles is blessing for a long and healthy life.

재료

주재료
소면 400g · 쇠고기육수 8C (양지편육 조리법 참조)

부재료
쇠고기 100g (우둔, 채 썰기) · 애호박 150g (채 썰기) · 달걀 1개 · 홍고추 1/2개

고기양념
간장 1T · 설탕 1/2T · 다진 마늘 1t · 다진 파 2t · 참기름 1t · 깨소금 1t · 후춧가루 약간

만드는 법

1. 소면 삶기
- 냄비에 물을 넉넉히 붓고 끓으면 국수를 펼쳐서 넣는다.
- 끓어오르면 찬물을 넣고 다시 끓인다.
- 국수가 익으면 찬물에 헹군다.
- 1인분씩 사리를 만들어 놓는다.

2. 육수 만들기
- 양지편육 조리법 참조

3. 쇠고기 양념하여 볶기
- 쇠고기는 고기양념을 넣고 5~10분간 재워둔다.
- 팬에 볶아낸다.

4. 애호박 볶기
- 소금에 호박을 10분간 절였다가 물기를 제거한다.
- 살짝 볶아낸다.

5. 지단 만들기
- 달걀을 황 · 백으로 나눈 뒤 지단을 부쳐 채 썬다.

6. 육수 끓이기
- 쇠고기 육수는 펄펄 끓여 국간장으로 색을 내고 소금으로 간한다.

7. 담기
- 그릇에 국수를 담고 고기, 호박나물, 황 · 백지단을 올린다.
- 뜨거운 육수를 부어낸다.

Ingredients

Main ingredients
Somyeon (wheat noodles) 400g · Beef Broth 8C (refer to Yangji-pyeonyuk recipe)

Sub-ingredients
Beef 100g (top round, julienne) · Green Squash 150g (julienne) · Egg 1ea · Red Chili 1/2ea (seeded, julienne)

Seasonings for beef
Soy Sauce 1T · Sugar 1/2T · Garlic 1t (minced)
Green Onion 2t (finely chopped) · Sesame Oil 1t
Sesame Seeds 1t (ground) · Black Pepper pinch (ground)

Steps

1. Boiling noodles
- Put noodles in enough boiling water.
- Once water boils again, add cold water.
- When noodles are cooked, drain and rinse thoroughly.
- Twist to shape noodles into rounds for individual serving size.

2. Making broth
- Refer to Yangji-pyeonyuk recipe.

3. Seasoning and stir-frying beef
- Marinate beef for 5~10 minutes.
- Stir-fry it.

4. Cooking green squash
- Sprinkle salt over green squash and set aside for 10 min.
- Squeeze out excess water.
- Stir-fry green squash.

5. Making egg jidan
- Separate egg yolk and egg white.
- Season with salt.
- Pan-fry egg yolk and egg white making thin layers and cool down.
- Julienne the egg Jidan.

6. Boiling beef broth
- When beef broth is boiling, add Gukganjang (soy sauce for soup) for color, add salt for taste.

7. Plating
- In a bowl, put noodles and top with beef, green squash, egg Jidan, red chili.
- Pour hot beef broth (from no. 6) over noodles.

귀띔 국수를 헹굴 때 계속 비비면서 끈기를 없애주어야 국수가 붙지 않는다.

Tips Rinse noodles in cold water rubbing continuously so that they won't stick together.

해물파전

해물파전은 한국식 팬케이크로 전채, 간식, 주음식으로 즐겨 먹는 음식이다.
서양인들은 해물파전을 피자에 빗대어 말하기도 한다.

Haemul-pajeon_ *Seafood and Scallion Pan-fry*

Haemul-pajeon is a Korean-style-pancake which is very popular as a main dish, appetizer or snack. It is often called Korean-style-pizza by foreigners.

재료

주재료
실파 120g (12cm 길이로 썰기) · 관자 30g (굵게 다지기) · 새우살 30g (내장 제거, 굵게 다지기) · 오징어 30g (굵게 다지기) · 조갯살 20g (굵게 다지기)

부재료
달걀 1개

반죽재료
밀가루 100g · 찹쌀가루 40g · 멥쌀가루 60g · 소금 1/2t

멸치육수
멸치 5개 · 다시마 (사방 10cm) 1장 · 물 2C → 1¼c

초간장
간장 2T · 물 2T · 식초 1/2T · 설탕 1/2T

만드는 법

1. 육수 내기
- 냄비에 물 2컵, 멸치, 다시마를 넣고 3분 정도 끓인다.

2. 반죽하기
- 반죽재료를 모두 섞어 육수를 넣고 반죽한다.

3. 파전 지지기
- 팬에 기름을 두르고 파전 반죽을 넣는다.
- 파를 가지런히 펴서 얹은 후 해물을 올린다.
- 파전 반죽을 조금 얹어 잘 눌러준다.
- 파전 위에 달걀 줄알을 친다.
- 밑부분이 익으면 뒤집어 익힌다.

4. 담아내기
- 큰 접시에 먹음직스럽게 담아낸다.
- 초간장과 같이 낸다.

Ingredients

Main ingredients
Scallion 120g (cut into 12cm length) · Scallop 30g (chopped) · Shrimp 30g (deveined, chopped) · Squid 30g (chopped) · Clam Meat 20g (chopped)

Sub-ingredients
Egg 1ea

Batter ingredients
Flour 100g · Glutinous Rice Flour 40g · Rice Flour 60g · Salt 1/2t

Dried anchovy broth
Dried Anchovy 5 heads · Kelp 1ea (10×10cm) · Water 2C → 1¼C

Dip
Soy Sauce 2T · Water 2T · Vinegar 1/2T · Sugar 1/2T

Steps

1. Making anchovy broth
- Boil 2 cups of water, anchovy and kelp for 3 minutes.

2. Making batter
- Mix all batter ingredients with broth (no. 1) to become thick.

3. Pan-frying
- Preheat a frying pan with oil, spread a ladle of batter on it.
- Put scallion on the batter and top with seafood.
- Spread more batter on and beaten egg over it.
- When bottom is cooked, turn over and cook thoroughly.

4. Plating
- Transfer to a big plate.
- Serve with dip.

귀띔 Tips

1. 파가 고루 익도록 파의 뿌리부분과 잎부분을 지그재그로 올려준다.
2. 전을 자주 뒤집지 말고 뚜껑을 덮어 속까지 잘 익혀야 한다.

1. When putting scallions, alternate scallions' green and white.
2. Cover frying pan with lid, when cooking Pa-jeon thoroughly.

양지편육

쇠고기나 돼지고기를 덩어리째 삶아 베보자기에 싸서 무거운 것으로 누른 뒤 얇게 저민 것을
편육이라 한다. 잔치음식을 만들 때 양지머리나 사태를 삶아서 국물은 국수장국의 국물로 쓰고
고기는 건져서 편육으로 쓴다.

Yangji-pyeonyuk_ *Boiled Brisket*

Pyeonyuk is slice of beef or pork which is boiled and pressed down with a heavy stone.
After boiling. the meat broth can be used for noodle soup.
Pyeonyuk and noodles are often served for Korean traditional feast or banquet.

재료

주재료
쇠고기 800g (양지머리) · 물 15C

향신채소
대파 1뿌리 · 마늘 10톨 · 소금 1T

초간장
간장 2T · 식초 1/2T · 설탕 1/2T · 물 2T · 잣가루 1t

만드는 법

1. 핏물 빼기
- 양지머리는 찬물에 1시간 정도 담가 핏물을 뺀다.

2. 삶기
- 냄비에 물 15C을 넣고 끓을 때 양지머리와 향신채소를 넣고 1시간 정도 삶는다.
- 떠오른 불순물을 걷어낸다.
- 소금을 넣고 10분 정도 더 끓인다.
- 고기를 건져 젖은 베보자기에 싸고 무거운 것으로 눌러 모양을 잡는다.

3. 담기
- 편육은 얇게 썰어 그릇에 담는다.
- 초간장을 만들어 같이 낸다.
** 삶은 물은 면보로 걸러 육수로 사용한다.

Ingredients

Main ingredients
Beef 800g (brisket) · Water 15C

Aromatic vegetables
Green Onion 1 stem · Garlic 10 cloves · Salt 1T

Vinegar soy sauce dip
Soy Sauce 2T · Vinegar 1/2T · Sugar 1/2T · Water 2T · Pine Nuts 1t (ground)

Steps

1. Blooding beef
- Soak brisket in cold water for an hour.

2. Boiling
- Add blooded brisket, aromatic vegetables and 15 cups of water in a pot. Boil for 1 hour.
- Skim off impurities.
- Add salt and bring stock to boil for 10 min.
- When beef is cooked, take chunks out of broth.
- Cover beef with clean cloth and shape beef by pressing down with heavy stone or objects.

3. Plating
- Slice beef and transfer to plate.
- Serve with vinegar soy sauce dip.
** Strain broth using clean cloth for clearer soup.

귀띔 고기가 무르게 익었을 때 소금을 넣어야 고기와 국물 맛이 좋아진다.
Tips To aid good quality broth add salt after meat is fully cooked.

타락죽, 닭잣죽무침, 콩자반, 나박김치

과거에는 우유가 귀한 식재료여서 우유로 죽을 끓여 만든
타락죽은 왕에게 진상했던 귀한 음식이었으나, 요즘은 몸이
허약하고 식욕이 없는 소아나 환자의 영양식으로 누구나
먹을 수 있는 약선음식이 되었다.

Tarak-juk, Dak-jatjuk-muchim, Kong-jaban, Nabak-kimchi

In the past, milk was a very rare ingredient in cooking. Therefore, the Juk (porridge) made of milk was considered a precious dish that was only served to Korean Kings. Yet, these days it is regarded as a common dish that is enjoyed as medicinal food. It is a nutritious meal for babies, patients, and people who are weak.

타락죽

'타락'은 궁중에서 쓰던 우유의 명칭이다.
타락죽은 쌀을 갈아서 우유를 부어 끓인 보양이 되는 죽으로 궁중에서는 시월 초하루부터 정월에
이르기까지 내의원에서 만들어 임금님께 올리던 보양죽이다.

Tarak-juk_ *Milk Porridge*

'Tarak' is another name for milk which was used in the Korean royal court.
Tarak-juk is made of ground rice boiled with milk. It was served to kings from October to January as a
healthy food by the royal clinic.

재료

주재료
쌀 1C · 물 3C · 우유 3C

양념
소금 약간

고명
잣 1/2t

만드는 법

1. 쌀 불리기
- 쌀은 뿌연 물이 나오지 않을 때까지 여러 번 씻는다.
- 씻은 쌀은 1시간 정도 물에 담가두었다가 건져 물기를 뺀다.

2. 쌀 갈기
- 믹서에 불린 쌀과 물 2C을 넣고 곱게 간다.

3. 죽 쑤기
- 냄비에 쌀 간 것과 물 1C을 붓고 나무주걱으로 저으면서 끓인다.
- 쌀이 말갛게 되고 되직해지면 불을 약하게 줄이고 우유를 조금씩 부어가며 잘 섞이도록 저어준다.

4. 담아내기
- 다 된 우유죽은 그릇에 담고 위에 잣을 올린다.
- 소금을 함께 곁들여낸다.

Ingredients

Main ingredients
Rice 1C · Water 3C · Milk 3C

Seasonings
Salt pinch

Garnish
Pine Nuts 1/2t

Steps

1. Soaking rice
- Wash rice until the water doesn't get milky.
- Soak rice in water for an hour, then drain.

2. Grinding rice
- Grind soaked rice with 2C of water using food processor.

3. Making porridge
- Boil no. 2 with 1C of water. Keep stirring using wooden spoon.
- When it turns translucent and thick, cook it over low heat adding milk little by little and stir well.

4. Plating
- Pour the porridge in a bowl and garnish with pine nuts.
- Serve with salt for seasoning.

귀띔
Tips

1. 우유와 물을 동량으로 하는 것이 기본이며 더 짙은 맛을 내려면 우유의 양을 늘린다.
2. 소금 대신 설탕으로 간을 맞추기도 한다.

1. Put equal amount of water and milk normally, but adding more milk would enrich flavor.
2. May season with sugar instead.

닭잣죽무침

닭잣죽무침은 기름기 없는 닭 가슴살을 잘게 찢은 것과 곱게 채 썬 채소를 잣죽에 버무린 음식이다.
닭의 담백함, 채소의 신선함과 잣죽의 부드러움을 즐길 수 있는 음식이다.

Dak-jatjuk-muchim_ *Chicken with Pine Nuts Porridge Salad*

This dish is made with shredded chicken breast and vegetables with pine nuts porridge. You can enjoy not only fresh vegetables but pine nuts porridge at the same time.

재료

주재료
닭 가슴살 200g

부재료
양파 80g (채 썰기) · 오이 80g (돌려깎아 채 썰기, 4cm 길이) ·
당근 40g (채 썰기, 4cm 길이) · 배 80g (채 썰기, 4cm 길이)

양념
닭고기 밑양념 · 소금 1t · 흰 후추 약간

향신채
대파 1대 · 마늘 5톨 · 생강 1톨 · 통후추 1t

잣죽 드레싱
쌀 1/2C · 잣 1/3C · 물 2½C · 소금 1½t

만드는 법

1. 닭 밑준비하기
- 냄비에 물을 넣고 닭 가슴살, 향신채를 넣어 삶는다.
- 닭 가슴살이 익으면 건져서 잘게 찢는다.
- 소금, 후추로 밑간한다.

2. 부재료 손질
- 채 썬 양파를 물에 담가 매운맛을 뺀 후 물기를 제거한다.

3. 잣죽 드레싱 만들기
- 불린 쌀은 물 1½C를 넣고 믹서에 갈아 체에 밭쳐, 쌀 물만 받아둔다.
- 잣은 물 1C를 넣고 믹서에 갈아서 체에 밭쳐, 잣 물만 받아둔다.
- 잣 물을 먼저 끓이다가 쌀 물을 넣어 된 죽을 쑨다.
- 잣죽에 소금 간을 하고 식힌다.

4. 버무리기
- 닭 가슴살과 부재료를 합한 후 잣죽 드레싱을 뿌려 가볍게 무쳐준다.

5. 담아내기
- 그릇에 담아낸다.

Ingredients

Main ingredients
Chicken Breasts 200g

Sub-ingredients
Onion 80g (julienne) · Cucumber 80g (use of peel of skin, 4cm length, julienne) · Carrot 40g (4cm length, julienne) · Korean Pear 80g (4cm length, julienne)

Seasonings
For Chicken · Salt 1t · White Pepper pinch

Aromatic Vegetables
Green Onion 1 stem · Garlic 5 cloves · Ginger 1ea · Black Pepper 1t (whole)

Pine nuts porridge dressing
Rice 1/2C · Pine Nuts 1/3C · Water 2½C · Salt 1½t

Steps

1. Preparing chicken
- In a pot, pour water, put chicken and aromatic vegetables. Bring to boil.
- Once chicken breast are cooked, take chicken breast out of water then tear in parts.
- Sprinkle salt and white pepper over chicken breast for taste.

2. Preparing sub-ingredients
- Put julienne onion in cold water to ooze out spicy taste. Tap them to dry out using kitchen towel.

3. Cooking porridge
- Put soaked rice and water into food processor and grind them, strain the water.
- Put pine nuts and water into food processor and grind them, strain the pine nuts water.
- Put the pine nuts water into a pot and bring it to boil then add the rice water.
- Add salt to season then let them cool down.

4. Tossing
- Gently toss all the ingredients with pine nuts porridge.

5. Plating
- Transfer to plate and serve.

귀띔　잣죽 대신 호두죽, 깨죽, 타락죽으로 대신할 수 있다.
Tips　Walnut porridge, sesame porridge, and milk porridge may be used as substitute.

콩자반

콩은 밭에서 나는 고기라 불릴 만큼 영양이 풍부한 완전식품이다.
콩자반은 콩장 또는 콩조림으로도 불리며 콩을 간장에 조려 짭조름하게 만든 반찬이다.

Kong-jaban_ *Black Bean Cooked in Sweet Soy Sauce*

Black bean is often called a nutritionally complete food; so, it is sometimes referred to as "the meat of
the field". Kong-jaban, which is also called "Kong-jang" or "Kong-jorim", is black beans boiled down in
soy sauce. It is a salty side dish in Korean meals.

재료

주재료
검정콩 불린 것 1C

조림장
간장 2T · 설탕 1½T · 콩 불린 물 1½C

양념
물엿 1T · 참기름 1T · 통깨 1T

만드는 법

1. 콩 준비하기
- 콩 불린 물에 콩을 넣고 중간불에서 끓인다.
- 물이 끓기 시작하면 약불로 줄이고 콩이 부드러워질 때까지 삶는다.

2. 조림장 만들기
- 분량의 양념을 넣어 조림장을 만든다.

3. 조리기
- 1/2분량의 조림장을 넣고 약한 불에서 조린다.
- 나머지 조림장을 넣고 국물이 자작할 정도로 조린다.
- 물엿, 참기름, 통깨를 넣고 마무리한다.

4. 담아내기
- 그릇에 담아낸다.

Ingredients

Main ingredients
Black Bean 1C (soaked)

Sub-ingredients
Soy Sauce 2T · Sugar 1½T · Bean soaked water 1½C

Seasonings
Corn Syrup 1T · Sesame Oil 1T · Sesame Seeds 1T

Steps

1. Preparing beans
- Cook the black bean in bean soaked water over medium heat.
- Once the water is boiled, lower the heat until the beans are soften.

2. Making sauce
- Combine all sub-ingredients well.

3. Cooking
- Add 1/2 of sauce over cooked bean and simmering over low heat.
- Pour the rest of sauce until sauce is nice and thick.
- At the last minute, add corn syrup, sesame oil and sesame seeds and stir throughly.

4. Plating
- Plate in a bowl.

귀띔
검정콩을 깨끗이 씻은 후 콩분량의 3~4배 물에 담가 5시간 또는 하룻밤 정도 불린다.

Tips
Wash dried black beans clearly and soak in 3~4 times of water for 5 hours or overnight.

나박김치

나박김치는 무, 배추를 주재료로 해서 국물이 흥건하면서도 맵지 않고 삼삼하게 담가 먹는 김치이다. 어느 계절에나 먹을 수 있으며 젓갈을 쓰지 않는 것이 특징이다.

Nabak-kimchi

__ *Watery Kimchi Made with Radish and Napa Cabbage*

Radish and napa cabbage are the main ingredients of Nabak-kimchi. This dish is served with lots of liquid, and does not have strong salty flavor or spicy taste. Korean people eat this dish through-out the year. The distinct feature of Nabak-kimchi is that fermented fish sauce is not used.

재료

주재료
무 500g (3×2.5×0.4cm 썰기) · 배추속대 300g (3×2.5cm 썰기)

소금물
소금 2T · 물 2C

부재료
미나리 50g (3cm 길이 썰기)
대파 (흰 부분) 20g (3cm 길이 고운체) · 마늘 30g (고운체) · 생강 10g (고운체) · 홍고추 1개 (3cm 길이 고운체)

김칫국
물 15C · 소금 4T · 설탕 1T · 고춧가루 3T

만드는 법

1. 무 손질하기
- 무는 껍질째 깨끗이 씻는다.

2. 절이기
- 무와 배추를 소금물에 10분간 절인다.
- 절인 배추와 무는 물에 한 번 헹구어 물기를 뺀다.

3. 재료 섞기
- 절인 배추와 무에 미나리를 제외한 모든 부재료를 섞는다.

4. 김칫국 만들기
- 물에 소금과 설탕을 넣고 완전히 녹인다.
- 면보에 고춧가루를 넣고 소금물에 살살 주물러 비비면서 붉은색을 낸다.

5. 담그기
- 3번 과정에 김칫국을 부어준다.
- 마지막에 미나리를 넣고 간을 맞춘다.

Ingredients

Main-ingredients
Radish 500g (3×2.5×0.4cm paysanne) · Heart of Napa Cabbage 300g (3×2.5cm paysanne)

Brine
Salt 2T · Water 2C

Sub-ingredients
Korean watercress 50g (3cm length) · Green Onion (white part only) 20g (3cm fine Julienne) · Garlic 30g (fine julienne) · Ginger 10g (fine Julienne) · Red Pepper 1ea (3cm fine Julienne)

Kimchi liquid
Water 15C · Salt 4T · Sugar 1T · Red Chili Powder 3T

Steps

1. Preparing radish
- Clean radish with skin, do not peel the radish skin.

2. Brining
- Soak radish and napa cabbage in salt water for 10 minutes.
- Rinse salted radish and cabbage with water and drain off all the extra water.

3. Mixing
- Mix the salted radish, napa cabbage and all the sub-ingredients except Korean watercress.

4. Making Kimchi liquid
- Dissolve salt and sugar with water.
- Wrap the red chili powder with sachet.
- Rub the sachet with salted water until it gets colored.

5. Making Nabak-kimchi
- Add Kimchi Liquid to no. 3.
- Add the Korean watercress and adjust the seasoning.

167

귀띔 사과나 배를 함께 나박썰어 넣어도 좋다.
Tips Slices of apple and Asian pear can be added to Nabak-kimchi.

달걀찜, 감자소고기조림, 김치찌개, 두부조림

찬바람 부는 겨울날, 김장김치에 돼지고기 숭덩숭덩 썰어 넣고 누구나
쉽게 끓여 먹을 수 있는 우리의 대표주자 음식인 김치찌개

Dalgyal-jjim, Gamja-sogogi-jorim, Kimchi-jjigae, Dubu-jorim

One cold winter day, I cut the pork into small pieces and mixed it with Kimjang-kimchi to make Kimchi-jjigae. It is one of the simplest dishes that Koreans can easily make and come up with.

김치찌개

한국은 주식이 밥이므로 국이나 찌개가 항상 기본으로 들어간다. 김치를 주재료로 끓이는
김치찌개는 한국인이 즐겨 먹는 찌개 중 하나이다.

Kimchi-jjigae_ *Spicy Stew Made with Kimchi*

Koreans usually eat rice—their staple food—with a soup or a stew. In Korean table setting,
Kimchi-jjigae is a representative stew, as the name suggests, it is made of Kimchi, and it is one of
Koreans' most favorite stews.

재료

주재료
배추김치 400g (소를 털어내고 3cm 길이로 썰기) • 돼지고기 100g (삼겹살, 3cm 길이로 썰기) • 양파 100g (0.6cm 두께로 채 썰기) • 대파 1대 (어슷썰기) • 홍고추 1개 (어슷썰기)

부재료
물 6C • 식용유 1T • 김칫국 약간

양념
고춧가루 1T • 다진 마늘 1t • 설탕 1t

만드는 법

1. 재료 손질하기
2. 찌개 끓이기
- 달군 냄비에 식용유를 넣고 돼지고기를 볶는다.
- 중불에서 김치를 넣고 기름이 배어 부드러워지면 찬물을 넣고 끓인다.
- 국물이 팔팔 끓으면 양파, 다진 마늘과 고춧가루, 설탕을 넣어 은근한 불에 끓인다.
- 마지막에 대파와 홍고추를 넣고 김칫국으로 간을 한 뒤, 다시 한 번 끓여낸다.

3. 그릇에 담아내기
- 전골그릇이나 수프그릇에 담아낸다.

Ingredients

Main ingredients
Kimchi 400g (fermented, stuffings removed, chopped into 3cm) • Pork 100g (belly, bite-sized into 3cm) • Onion 100g (0.6cm, julienne) • Green Onion 1ea (bias-cut) • Red Chili 1ea (bias-cut)

Sub-ingredients
Water 6C • Cooking Oil 1T • Juice of Kimchi

Seasonings
Red Chili Powder 1T • Garlic 1t (minced) • Sugar 1t

Steps

1. Preparing ingredients
2. Boiling
- Preheat a pan with oil and stir—fry pork.
- Over medium heat, add Kimchi and cook to soften.
- Add cold water and bring to boil.
- When boiling, add onion, garlic, red chili powder and sugar.
- Reduce heat and simmer.
- When almost finished, add green onion and red chili.
- Allow to boil for seconds.

3. Plating
- Transfer to a stew or soup bowl to serve.

달걀찜

달걀은 영양소가 풍부한 식품으로 값이 싸고 생산량이 풍부하며 누구나 손쉽게 구할 수 있는
식재료이다. 달걀찜은 대표적인 달걀요리로 부드럽고 담백한 맛으로 대부분의 사람들이 선호하는
음식이다.

Dalgyal-jjim_ *Steamed Egg*

Egg (Dalgyal in Korean) is well-known for its perfect nutritional content and availability in low price.
And it is cooked in various ways, as well. This Dalgyal-jjim is a representative steamed or boiled egg
dish in Korea. Everyone likes its soft texture and mild flavor.

재료

주재료
달걀 4개

부재료
물 2C (달걀 양의 2배) • 새우젓국물 1T • 참기름 1t • 소금 약간

고명
실고추 (1cm 길이로 썰기) • 석이버섯 (가늘게 채 썰기, 참기름, 소금으로 볶기) • 실파 (파란 부분 가늘게 어슷썰기) 1줄기

만드는 법

1. 달걀과 육수 준비하기
- 달걀을 잘 풀어준다.
- 새우젓국물, 참기름, 소금을 넣고 약 3분간 둔다.
- 물을 섞어 체에 내리고, 그릇에 담는다.

2. 찜하기
- 냄비에 면보를 놓고 물을 붓는다.
- 1번의 찜그릇을 넣고 뚜껑을 덮어 끓인다.
- 끓기 시작하면 약불로 줄이고 10~15분 정도 찐다.

3. 고명 올리기
- 고명을 올리고 1분 정도 다시 찐다.

Ingredients

Main ingredients
Egg 4ea

Sub-ingredients
Water 2C (twice as much as egg) • Salted Shrimp Juice 1T • Sesame Oil 1t • Salt pinch

Garnish
Silgochu (thinly cut dried red chili, 1cm length) • Stone Ear Mushroom (julienne, stir-fried with sesame oil and salt) • Scallion (thinly bias-cut, green part only) 1 stem

Steps

1. Preparing egg and broth
- Beat eggs well.
- Add salted shrimp juice, sesame oil and salt, set aside for about 3 minutes.
- Mix water and strain using a sieve.
- Transfer to a bowl to steam.

2. Steaming
- Spread cotton cloth in a pot and add water.
- Place steam bowl on. Cover with a lid and steam it.
- When the water starts to boil, reduce the heat into low and steam for 10~15 min.

3. Garnishing
- Put the garnish on, cover and steam for a min.

173

감자소고기조림

감자소고기조림은 감자와 소고기를 조림장에 조린 것이다.
주로 밥과 함께 곁들여 먹는 반찬이다.

Gamja-sogogi-jorim

__ *Braised Potato and Beef with Sweet Soy Sauce*

Gamja-sogogi-jorim is potato and beef braised in soy sauce.
It is usually served as a side dish for cooked rice.

재료

주재료
감자 5개 (약 700g)

부재료
쇠고기 100g (납작하게 저미기) · 마늘 3톨 (편썰기) · 파 2개 (어슷썰기) · 홍고추 1/2개 (어슷썰기) · 풋고추 1/2개 (어슷썰기)

양념
쇠고기양념 · 간장 1T · 설탕 1/2T · 다진 마늘 1t · 다진 파 2t · 참기름 1t
조림양념 · 물 3C · 설탕 2T · 진간장 1T · 참기름 1/2t · 물엿 1T · 소금 · 후춧가루 약간

만드는 법

1. 감자 손질하기
- 감자의 크기가 작은 것은 둘로, 큰 것은 (십자로 잘라) 네 조각으로 자른다.
- 자른 감자는 끓는 물에 소금을 약간 넣고 삶아낸다.

2. 볶기
- 양념한 쇠고기를 냄비에 넣고 살짝 볶는다.
- 기름과 감자를 넣고 볶아 기름을 충분히 먹인다.

3. 감자 조리기
- 볶은 쇠고기와 감자에 양념장을 붓고 끓인다.
- 물기가 반으로 줄어들면 불을 줄이고 수분이 거의 없을 때까지 푹 조린다.
- 마늘, 파, 홍고추, 풋고추와 참기름, 물엿을 넣은 뒤 고루 저어 마무리한다.
- 소금, 후추로 간을 맞춘다.

Ingredients

Main ingredients
Potato 5ea (about 700g)

Sub-ingredients
Beef 100g (sliced, bite-sized) · Garlic 3 cloves (sliced) · Green Onion 2ea (bias-cut) · Red Chili 1/2ea (bias-cut) · Green Chili 1/2ea (bias-cut)

Seasonings
For beef marinating · Soy Sauce 1T · Sugar 1/2T · Garlic 1t (minced) · Green Onion 2t (finely chopped) · Sesame Oil 1t
For braising · Water 3C · Sugar 2T · Soy Sauce 1T · Sesame Oil 1/2t · Corn Syrup 1T · Salt & Black Pepper pinch

Steps

1. Preparing potato
- Cut a potato into 2 or 4 pieces.
- Boil the potatoes in salted water.

2. Stir-frying
- Slightly stir-fry the seasoned beef on low heat.
- Stir-fry the potatoes until the potatoes absorb the oil.

3. Braising
- Braise the beef and potatoes with seasoning sauce.
- When sauce is reduced to half, lower the heat and braise until the liquid has evaporated.
- At the last minute, add garlic, green onion, red chili, green chili, sesame oil, corn syrup and mix well.
- Add salt and black pepper to taste.

175

두부조림

두부조림은 두부를 썰어 전으로 지진 다음 간장에 조려낸 음식으로 반상차림에 잘 어울리는
반찬이다.
두부는 콩을 갈아 만들어 소화율이 높으며 양질의 단백질이 함유된 식품이다.

Dubu-jorim

__ *Pan-fried Dubu Simmered in Soy Sauce*

In Dubu-jorim, dubu is pan-fried and boiled down with soy sauce. Dubu is made of ground bean. Dubu
is easy to digest and contains healthy proteins.

재료

주재료
두부 400g (5×4×1.5cm 썰기) • 소금 약간

부재료
대파 1/2뿌리 (3cm 채 썰기)

양념
진간장 3T • 설탕 1T • 다진 파 2t • 다진 마늘 1t • 고춧가루 1t • 깨소금 1t • 참기름 1t • 후추 약간 • 물 1/2C

만드는 법

1. 두부 손질하기
- 두부는 크기대로 썰어 소금을 뿌려 밑간한다.

2. 양념장 만들기
- 그릇에 양념을 모두 넣고 섞는다.

3. 두부 지지기
- 밑간한 두부는 물기를 제거한다.
- 팬에 기름을 두르고 두부를 앞, 뒤로 노릇하게 지져낸다.

4. 두부 조리기
- 두부에 양념장, 물을 넣고 중불로 줄인다.
- 두부가 어느 정도 조려지면 파채를 얹는다.

5. 담아내기
- 완성된 두부를 접시에 담아낸다.

Ingredients

Main ingredients
Dubu 400g (cut in 5×4×1.5cm) • Salt pinch

Sub-ingredients
Green Onion 1/2ea (3cm julienne)

Seasonings
Soy Sauce 3T • Sugar 1T • Green Onion 2t (finely chopped) • Garlic 1t (minced) • Red Chili Powder 1t • Sesame Seeds 1t • Sesame Oil 1t • Black Pepper pinch • Water 1/2C

Steps

1. Preparation
- Slice dubu and season with salt.

2. Seasoning
- Mix all seasoning ingredients.

3. Pan-frying
- Remove excess water from dubu.
- Pan-fry dubu until golden brown on both side.

4. Simmering
- Place the dubu in a pan with water and seasoning and simmer over medium heat.
- Garnish with julienned green onion.

5. Plating
- Transfer to a plate.

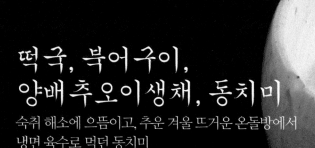

떡국, 북어구이,
양배추오이생채, 동치미

숙취 해소에 으뜸이고, 추운 겨울 뜨거운 온돌방에서
냉면 육수로 먹던 동치미

Ttoek-guk, Bugeo-gui,
Yangbachu-oi-saengchae,
Dongchimi

I used to sit on the Ondolbang, the heated floor, and enjoyed
Naengmyeon (cold noodles with Dongchimi) during the freezing
winter. Dongchimi was very effective in reducing hangovers as it is
used as a broth for Naengmyeon.

떡국

정월 초하루에 떡국을 먹어야 한 살 더 먹었다고 할 정도로 나이를 떡국 그릇 수에 비유하기도 한다. 가래떡은 희고 길어 순수와 장수를 의미하고 떡국 떡은 모양이 돈과 같다 하여 부를 상징하는 의미를 갖는다. 새해 차례음식으로 밥 대신 올리고 어른들께 세배 올린 후 세찬 상으로 낸다.

Tteok-guk_ *Rice Cake Soup*

Koreans eat this Rice Cake Soup on the New Year's Day, with Tteok-guk symbolizing becoming a year order. Garae-tteok, which is a main ingredient of this dish, represents longlife and wealth by its length and coin shape. Tteok-guk is prepared for the New Year's memorial service and shared with families.

재료

주재료
떡국 떡 500g

부재료
소고기 (양지) 200g (한입 크기로 썰기) · 달걀 2개 · 대파 50g (어슷썰기) · 국간장 1t · 소금 약간 · 물 6C

양념
고기양념 · 국간장 1/2T · 다진 파 2t · 다진 마늘 1t · 참기름 1t · 후추 약간

만드는 법

1. 국물내기
- 소고기에 밑간하기
- 달군 냄비에 양념한 쇠고기를 넣고 볶다가 물을 넣고 중불에서 끓인다.

2. 떡국 끓이기
- 국물이 끓으면 국간장과 소금으로 간을 하고 가래떡을 넣는다.
- 떡이 물 위로 떠오르면 준비된 달걀의 줄알을 친 후 달걀이 익으면 파를 넣고 불을 끈다.

Ingredients

Main ingredients
Stick Rice Cake 500g (coin shape sliced)

Sub-ingredients
Beef 200g (Brisket, bite-sized) · Egg 2ea · Green Onion 50g (bias-cut) · Gukganjang 1t · Salt to taste · Water 6C

Seasonings
For beef · Gukganjang 1/2T · Green Onion 2t (finely chopped) · Garlic 1t (minced) · Sesame Oil 1t · Black Pepper pinch

Steps

1. Making broth
- Marinate brisket with seasonings.
- Stir-fry brisket in a pre-heated pan and bring to boil by adding water over medium heat.

2. Making tteok-guk
- When the broth starts boiling, season with gukganjang and salt. And add stick rice cake.
- When garae-tteok floats, add beaten egg stirring gently.
- When egg is cooked, add green onion and turn off heat.

181

귀띔
Tips

1. 식성에 따라 후추나 채 썬 김을 넣어 먹기도 한다.
2. 사골과 양지 육수가 대표적이지만 조개나 멸치 등 다양한 육수를 이용해도 좋다.
3. 떡국에 만두를 넣어 먹으면 별미다.

1. Pepper and crushed and dried seaweed can be added.
2. Clam, anchovy or other various type of broth can substitute a meat-based broth.
3. Adding dumplings to in the soup gives more flavor.

설날 떡국상차림

음력 정월 초하루가 되면 그동안 떨어져 있던 가족 및 친지들이
한자리에 모여 차례를 지낸다. 부모와 웃어른들에게 세배를 하고
덕담을 주고받으며 세찬을 한다.

Table with Ttoek-guk on the
New Year's Day

On the first day of the year in the Lunar calendar, families and cousins
who live far from one another get together and have a memorial service
for their ancestors. Young ones perform the New Year's bows to elders and
share well-wishing remarks and food.

양배추오이생채

양배추와 오이를 채 썰어 갖은 양념하여 새콤하게 무쳐 먹는 생채이다.
'생채'는 날채소를 채 썰어 무친 나물이다.

Yangbaechu-oi-saengchae

_ *Cabbage and Cucumber with Spicy Dressing*

Saengchae means freshly-sliced vegetables tossed with spicy dressing.
Yangbaechu-oi-saengchae is made of cabbage (yangbaechu) and cucumber (oi).

재료

주재료

양배추 100g (2×4cm 편썰기) • 오이 100g (2×4cm 편썰기)

양념

간장 1/2T • 고춧가루 1/2T • 설탕 1/2T • 식초 1/2T • 액젓
1t • 다진 마늘 1t • 참기름 1t • 깨소금 1t

만드는 법

1. 양념장 만들기
 -모든 재료를 잘 섞는다.
2. 무치기
 -썰어둔 오이와 양배추에 양념장을 넣어 버무린다.
3. 접시에 담기
 -접시에 소담스럽게 담는다.

Ingredients

Main ingredients

Cabbage 100g (2×4cm sliced) • Cucumber 100g (2×4cm sliced)

Spicy dressing

Soy Sauce 1/2T • Red Chili Powder 1/2T • Sugar 1/2T • Vinegar
1/2T • Fish Sauce 1t • Garlic 1t (minced) • Sesame Oil
1t • Sesame Seeds 1t (ground)

Steps

1. Making spicy dressing
 -Mix all dressing ingredients well.
2. Mixing/ tossing
 -Gently toss prepared cabbage and cucumber with spicy dressing.
3. Plating
 -Transfer to a plate.

185

북어구이

북어는 명태를 얼리면서 말린 것으로 살이 부풀어서 마치 더덕처럼 된 것을 더덕북어라 하여
최고로 친다. 말린 것이라 상하지 않으므로 실온에 두고 아무 때나 꺼내 쓸 수 있어 편리하다.
껍질째 매콤하게 양념한 북어구이는 일품음식으로도 손색이 없다.

Bugeo-gui

__ *Pan-fried Dried Pollack with Red Chili Paste Sauce*

Bugeo is made of walleye pollack, which is dried while freezing and its flesh swells like deodeok and
is called deodeokbugeo, known to be one of the best dried pollacks. Bugeo does not get rotten at room
temperature as it is dry, so it is very convenient to cook anytime.

재료

주재료
북어포 2마리 (껍질 있는 것)

양념
고추장 3T · 물 2T · 간장 1t · 설탕 1/2T · 고춧가루 1T ·
맛술 1T · 물엿 1T · 마늘 1T · 깨소금 1T · 참기름 1T

만드는 법

1. 북어 손질하기
- 북어포를 두어 번 헹구어 물기를 꼭 눌러서 머리, 꼬리,
  지느러미를 잘라내고 가시를 발라낸 후 껍질 쪽에 2cm
  간격으로 얕게 칼집을 넣는다.

2. 양념장 바르기
- 분량의 구이 양념을 한데 섞어 양념장을 만든 뒤 북어에
  고루 바르고 10여 분간 재워 놓는다.

3. 북어 굽기
- 팬을 달구어 기름을 넉넉히 두르고 북어의 살 부분을 먼
  저 익힌 뒤, 뒤집어 다시 한 번 양념장을 발라 굽는다.

Ingredients

Main ingredients
Dried Pollack 2 heads (skin-on)

Seasonings
Red Chili Paste 3T · Water 2T · Soy Sauce 1t · Sugar 1/2T · Red
Chili Powder 1T · Cooking Wine 1T · Corn Syrup 1T · Garlic
1T · Sesame Seeds 1T (ground) · Sesame Oil 1T

Steps

1. Trimming dried pollack
- Wash the dried pollack twice or more and tap—dry to get rid of
  the remaining water.
- Cut off the head, tail and fin and remove bones.
- Give light slits at intervals of 2cm on skin.

2. Spreading the seasoning
- Seasonings to spread over the dried pollack and leave it for about
  10 minutes.

3. Pan-frying
- Heat the pan and drizzle little oil.
- Cook the inner flesh side first, turn over and spread the seasoning
  once more.

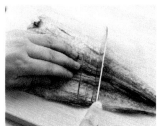

귀띔
Tips

1. 말린 북어를 불릴 때는 반드시 찬물에 불리는 것이 중요하다.
   뜨겁거나 미지근한 물에 불리면 살이 풀어지고 영양손실도 많다.
2. 껍질 쪽에 칼집을 넣어주어야 구울 때 오그라들지 않는다.

1. It is very important to soak dried pollack in cold water.
   If soaked in hot water, it loses flavor, texture and nutrition.
2. Giving slits on the skin prevents the meat from shrinking.

동치미

물김치의 일종으로 무와 배추, 배 등을 넣고 익힌 전통적으로 겨울에 담가 먹는 김치이다.

Dongchimi

__ *Watery Radish Kimchi*

Dongchimi is not only a simple watery radish Kimchi but also the best way to enjoy real fermented flavor and refreshing taste.

재료

주재료
무 3개 (5.5×1×1cm, 막대모양) · 소금 2T (호렴)

부재료
삭힌 고추 5개 · 배 1개 · 파 3대 · 다진 마늘 3T · 생강즙
1½T · 무즙 500g (고운체에 거르기) · 쪽파 30g

양념
물 30C · 소금 1C · 시럽 1C (물 1C+설탕 1C)

만드는 법

1. 재료 준비하기
- 남은 무 자투리는 믹서에 곱게 갈아둔다.
- 자른 무는 소금에 절인다.
- 배는 껍질째 씻어 반쪽은 4등분해서 넣고, 나머지 반은 갈아 즙으로 만든다.
- 쪽파는 소금에 살짝 절여 매듭짓는다.

2. 시럽 만들기
- 냄비에 물과 설탕을 넣고 끓인다.

3. 동치미 국물 만들기
- 물에 소금, 시럽, 무즙, 배즙, 생강즙을 넣고 고루 섞어준다.

4. 동치미 만들기
- 항아리에 무, 배, 고추를 담는다.
- 주머니에 파, 마늘을 넣어 항아리의 중간쯤에 담아둔다.
- 준비해 둔 동치미 국물을 부어준다.

5. 보관하기
- 하룻밤 상온에 두었다가 냉장보관한다.

Ingredients

Main ingredients
Radish 3ea (skinned, 5.5×1×1cm stick shape) · Salt 2T (sea salt)

Sub-ingredients
Green Chili 5ea (fermented) · Korean Pear 1ea · Green Onion 3 stems · Garlic 3T (minced) · Ginger Juice 1½T · Radish Juice 500g · Scallion 30g

Seasonings
Water 30C · Salt 1C · Sugar Syrup 1C (water 1C+sugar 1C)

Steps

1. Preparing ingredients
- Blend trimmed pieces of radish.
- Salting the stick shaped radish.
- Quarter a half of pear with the skin on, and make juice another half.
- Slightly salt in the scallions and tie.

2. Making syrup
- Add water and sugar in a small pot and bring to boil.

3. Making dongchimi liquid
- Mix water, salt, syrup, pear juice, ginger juice and radish juice in a bowl.

4. Making dongchimi
- Put radish, pear and green chili in a jar.
- Make a cloth pocket with green onion and garlic in. Keep it in the middle of the jar.
- Pour dongchimi liquid in the jar.

5. Storing
- Keep the jar in room temperature for one night, then refrigerate.

귀띔 Tips

1. 무즙은 고운체에 걸러 넣어야 국물이 깨끗한 맛이 난다.
2. 삭힌 고추가 없을 때는 청양고추를 소금물에 절였다가 쓰면 풋내가 덜하다.

1. Use a fine sieve to strain radish for refreshing taste.
2. In case you don't have any fermented chili, use salted Cheongyang chili (a kind of very spicy chili in Korea) to prevent the grassy smell.

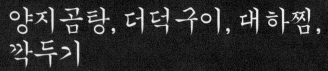

양지곰탕, 더덕구이, 대하찜, 깍두기

곰탕은 본래 소고기와 소 내장을 진하게 고아 만드는 것이다.
양질의 단백질 급원으로 체력을 증진시키기 위한 음식으로 임금님도
즐겨 드셨던 보양음식이다.

Yangji-gomtang, Deodeok-gui, Daeha-jjim, Kkakdugi

Gomtang is a soup made with thoroughly boiled beef and cow offal. As a source of high-quality protein, Korean Kings used to enjoy this soup as an invigorating food.

양지곰탕

양지곰탕은 양지에 물을 붓고 오래 끓인 국이다.
탕이나 국에는 밥이 같이 제공되고 한국인들은 국이나 탕에 밥을 말아 먹는 것을 즐긴다.

Yangji-gomtang_ *Clear Brisket Soup*

Yangji-gomtang is a soup with beef brisket boiled for long time. Most guk or tang is served with a bowl of rice. Koreans enjoy this dish by putting rice into soup and mixing it together.

재료

주재료
쇠고기 1kg (양지 덩어리) · 무 500g · 물 5L · 국간장 약간

향신채소
대파 1뿌리 · 마늘 10톨

고기와 무 양념
국간장 2T · 다진 마늘 2T · 다진 파 4T · 참기름 2T · 소금 2t · 후추 약간

초간장
간장 2T · 식초 2t · 물 2t · 잣가루 1t

만드는 법

1. 핏물 제거하기
-양지의 기름을 뗀다.
-찬물에 1시간 이상 담가 핏물을 제거한다.

2. 끓이기
-찬물에 고기와 무, 향신채소를 넣고 센 불에서 끓인다.
-끓기 시작하면 불을 줄여 1시간 정도 더 끓인다.

3. 고기와 무 양념하기
-양지는 고깃결 반대로 얇팍하게 썬다.
-무는 2×3×0.2cm 크기로 도톰하게 썬다.
-고기와 무에 쓸 양념장을 만든다.
-고기양념장 만든 것에 썬 고기와 무를 양념한다.

4. 기름 걷어내기
-국물은 식힌 뒤 기름을 걷어낸다.

5. 끓여 간 맞추기
-기름기 걷어낸 국물을 다시 끓인다.
-양념한 고기와 무를 넣는다.
-국간장으로 간한다.

Ingredients

Main ingredients
Beef 1kg (brisket) · Radish 500g · Water 5L · Gukganjang to taste

Aromatic vegetables
Green Onion 1 stem · Garlic 10 cloves

Seasonings for beef and radish
Gukganjang 2T · Garlic 2T (minced) · Green Onion 4T (finely chopped) · Sesame Oil 2T · Salt 2t · Black Pepper pinch

Vinegar soy sauce dip
Soy Sauce 2T · Vinegar 2t · Water 2t · Pine Nuts 1t (ground)

Steps

1. Blooding beef
-Soak brisket chunks in cold water for 1 hour.

2. Boiling
-Add the blooded brisket, radish and water with aromatic vegetables and boil over high heat.
-When starting to boil, reduce heat into low and simmer for 1 hours.

3. Seasoning beef and radish
-Take beef and radish out of stock.
-Slice cooked beef brisket.
-Cut cooked radish into 2×3×0.2cm sized pieces.
-Mix all the seasonings in a bowl.
-Add beef and radish in mixed seasoning and toss well.

4. Skimming
-Chill the broth and remove the fat on surface.

5. Boiling and seasoning
-Bring the skimmed broth to boil.
-Add seasoned beef and radish.
-Season with gukganjang to taste.

193

대하찜

대하찜은 궁중에서 먹던 대표적인 해물음식으로 껍질 벗긴 새우살이 들어가 먹기가 쉽고 채소와
고기가 잣집으로 어우러져 그 맛이 담백하고 고소하다.

Daeha-jjim

__ *Cooked Prawn Tossed with Vegetables in Pine Nuts Sauce*

Daeha-jjim is the most representative seafood in Korean royal cuisine. Steamed prawns vegetables, and
beef are mixed with pine nuts sauce.
Since this dish uses only the flesh of prawn it is easy to eat. The pine nuts sauce gives it a rich, nutty
taste.

재료

주재료
대하 4마리 (내장 제거)

향신재료
생강 10g (편썰기) · 대파 30g · 청주 1T · 물 1/2C

부재료
쇠고기 70g (사태살, 삶은 후 납작하게 썰기) · 오이 1개 (반 가른 후 어슷썰기) · 죽순 50g (빗살모양으로 얇게 썰기) · 소금 약간 · 흰 후춧가루 약간

잣집
잣가루 4T · 대하국물 3T · 참기름 1t · 소금 2/3t · 흰 후춧가루 약간

만드는 법

1. 대하 찌기
- 찜통에 분량의 향량재료를 넣고 7~8분 정도 찐다.
- 이때 밑으로 흘러내린 대하국물은 받아둔다.
- 쪄진 대하는 머리, 꼬리를 제거하고 껍질을 벗겨낸다.
- 새우살은 3cm 크기로 어슷썬다.

2. 부재료 준비하기
- 오이는 소금에 10분간 절인 뒤 물기를 짜고 살짝 볶아낸다.
- 죽순은 소금, 후추를 뿌려 밑간한 뒤 볶아낸다.

3. 잣집 만들기
- 모든 양념을 넣고 잘 저어 잣집을 만든다.

4. 무치기
- 새우와 부재료에 잣집을 넣어 가볍게 무친다.

5. 그릇에 담기
- 접시에 담는다.

Ingredients

Main ingredients
Prawn 4ea (deveined)

Aromatic ingredients
Ginger 10g (sliced) · Green Onion 30g · Rice Wine 1T · Water 1/2C

Sub-ingredients
Beef 70g (shank, boiled, sliced) · Cucumber 1ea (halved, bias-cut) · Bamboo Shoot 50g (boiled, sliced (comb shape)) · Salt pinch · White Pepper Powder pinch

Pine nuts sauce
Pine Nuts 4T (ground) · Prawn Stock 3T · Sesame Oil 1t · Salt 2/3t · White Pepper Powder pinch

Steps

1. Preparing prawn
- Steam prawns with aromatic ingredients for 7~8 min.
- Collect prawn stock for later use.
- Remove head, shell and tail.
- Slice prawn diagonally into about 3cm pieces.

2. Preparing sub-ingredients
- Salting the cucumber for 10 min. then squeeze out excess water, stir-fry the cucumber.
- Season the bamboo shoots with salt and white pepper and stir-fry.

3. Making pine nuts sauce
- Mix all ingredients for pine nuts sauce.

4. Tossing
- Add prawn, sub-ingredients and pine nuts sauce, gently toss them.

5. Plating
- Transfer to a plate.

외국인도 빠져드는 한국밥상

195

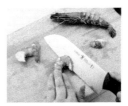

귀띔 잣집을 만들 때 크림상태로 으깨듯이 한참 젓는다.

Tips Whip the pine nuts sauce until it becomes creamy.

깍두기

무는 사철 흔히 볼 수 있는 채소지만 김장철에 나오는 무가 특히 달고 단단하여 저장성이 있는 깍두기 등의 무 반찬을 만들기에 가장 알맞다. 무는 위에서 소화를 돕기 때문에 고기를 주로 한 국이나 탕과 함께 먹으면 좋다.

Kkakdugi_ *Radish Kimchi*

Kkakdugi means cutting objects into cubes.
Radish is found all year round in Korea, but autumn radish is especially good and tasty. Radish helps digestion in the stomach, so Kkakdugi goes well with all kinds of tang or guk that are mostly meat based.

재료

주재료
무 2kg (2.5×2×1.5cm 깍둑썰기) · 호렴 (소금 4T, 물 1C)

부재료
쪽파 30g (3cm 길이로 썰기) · 미나리 30g (3cm 길이로 썰기) · 대파 70g (어슷썰기)

양념
고춧가루 5T · 새우젓 2T · 다진 마늘 1/2T · 다진 생강 1t · 소금 1t · 설탕 1T

만드는 법

1. 절이기
- 깍둑썰기한 무는 소금에 2시간 정도 절인다.
- 절인 무를 헹구고 물기를 뺀다.

2. 고춧물 들이기
- 깍둑썬 무에 고춧가루를 넣고 고춧물을 들인다.

3. 양념 만들기
- 고춧가루를 뺀 나머지 양념을 섞어 양념을 만든다.

4. 버무리기
- 고춧물 들인 무에 양념을 넣고 버무린다.
- 부재료를 모두 넣고 가볍게 섞는다.

5. 숙성하기
- 실온에서 하루 정도 두었다가 냉장보관한다.

Ingredients

Main ingredients
Radish 2kg (2.5×2×1.5cm cubes)
Brine (sea salt 4T, water 1C)

Sub-ingredients
Scallion 30g (cut into 3cm length) · Minari (Korean watercress) 30g (cut into 3cm length) · Green Onion 70g (bias-cut)

Seasonings
Red Chili Powder 5T · Salted Shrimp 2T (finely chopped) · Garlic 1/2T (minced) · Ginger 1t (minced) · Salt 1t · Sugar 1T

Steps

1. Salting radish
- Salting cubed radish for about 2 hours.
- Rinse the salted radish and drain.

2. Coloring radish
- Mix salted radish and red chili powder well and set aside.

3. Making seasonings
- Mix all seasonings except red chili powder.

4. Mixing
- Add salted and colored radish into seasonings and mix well.
- Add all sub-ingredients and gently toss them.

5. Fermenting and storing
- Ferment in room temperature for one day, then refrigerate.

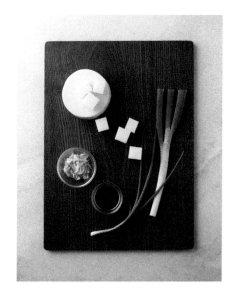

귀띔 무가 맛없는 여름에는 설탕과 소금을 넣어 절인 후 나오는 물은 버리고 사용한다.

Tips During Summer, add sugar first prior to salting radish since radish is less sweet.

더덕구이

더덕은 특유의 쌉싸름한 맛과 향긋한 향으로 많은 사랑을 받는다. 더덕은 감기와 기관지염에 매우 좋으며 폐와 비장, 신장을 튼튼하게 해주는 식품이다. 양념을 발라 구운 더덕구이가 더덕음식의 으뜸이라 할 수 있다.

Deodeok-gui_ *Grilled or Pan-fried Deodeok*

Deodeok. has a unique. strong scent and a bittersweet taste. Deodeok is considered to be a herbal medicine and especially good for a cold and bronchitis. This spicily seasoned 'Deodeok-gui' is the best dish among deodeok dishes.

재료

주재료
더덕 200g (껍질 벗겨서 반 가름) · 소금물 (소금 1t + 물 1C)

유장
참기름 2T · 간장 2t

양념
고추장 2T · 고춧가루 2t · 설탕 2t · 다진 마늘 2t · 참기름
1t · 물 1T · 물엿 2t · 깨소금 1/2t

만드는 법

1. 더덕 손질하기
- 소금물에 5분 정도 담갔다가 꺼내서 물기를 없앤다.
- 밀대로 두드려 넓게 편다.

2. 유장 처리
- 유장을 더덕에 바른 후 살짝 구워낸다.

3. 양념장 만들기
- 양념재료를 모두 섞어 양념장을 만든다.

4. 양념장 발라 굽기
- 양념장을 더덕에 바른 후 팬에 구워낸다.

5. 담아내기
- 그릇에 가지런히 담아낸다.

Ingredients

Main ingredients
Deodeok 200g (peeled, halve) · Brine (salt 1t + water 1C)

Yujang (oil sauce)
Sesame Oil 2T · Soy Sauce 2t

Seasonings (gochujang sauce)
Red Chili Paste 2T · Red Chili Powder 2t · Sugar 2t · Garlic 2t
(minced) · Sesame Oil 1t · Water 1T · Corn Syrup 2t · Sesame
Seeds 1/2t (ground)

Steps

1. Preparing deodeok
- Soak deodeok in brine for 5 minutes and tap with cotton cloth to
  remove water.
- Pound gently and flatten using rolling pin.

2. Seasoning with yujang
- Coat deodeok with yujang and slightly cook on a pan.

3. Making gochujang sauce
- Mix all seasonings.

4. Marinating and cooking
- Marinate deodeok (no. 2) in gochujang sauce.
- Cook over low heat.

5. Plating
- Transfer to a flat plate.

김치, 보쌈, 보쌈무생채, 우거지된장국

한국의 '김치와 김장'문화는 유네스코 인류무형문화유산에 등재되었으며, 긴 동절기 기간 동안 김치를 먹기 위해 준비하는 것이 김장이다. 김장을 할 때에는 품앗이로 이웃 간에 나눔의 정신을 실천하고, 그들 사이에 연대감과 소속감을 부여하는 데 그 의미가 있다.

Kimchi, Bossam, Bossam-mu-saengchae, Ugeoji-doenjang-guk

Korea's "Kimchi and Kimjang" culture means preparing Kimchi to enjoy during the winter time. UNESCO recently recognized and inscribed this onto its list of intangible cultural heritage items. Kimjang promotes the spirit of sharing among neighbors in Korea, while encouraging solidarity and providing people with a sense of identity and belongings.

김치

김치는 한국의 대표적인 전통발효식품으로 쌀 위주인 식생활에서 가장 중요한 부식의
하나이다. 예부터 채소류가 재배되지 않는 겨울철에 주로 많이 담근 저장식품으로 다양한
영양공급원이었다.

Kimchi_ *Napa Cabbage Kimchi*

Kimchi is the most representative fermented dish in Korea. It plays a key role in side dishes, as Koreans
live on rice.
Since it was quite difficult to harvest vegetables in the winter, this long-storable fermented dish became
the most important source of minerals and amino acids.

재료

주재료
배추 3.4kg · 굵은소금 1C

부재료
무 1.6kg (채 썰기) · 쪽파 100g (4cm 길이로 썰기) · 미나리
150g (4cm 길이로 썰기) · 찹쌀가루 3T · 물 1½C

양념
고춧가루 1C (1/2C 무에 사용, 1/2C 찹쌀풀에 사용) · 액젓 3T ·
새우젓 3T (곱게 다지기) · 설탕 2T · 다진 마늘 4T · 다진
생강 2t · 소금 2T

만드는 법

1. 배추 절이기
- 배추의 겉잎을 떼어내고 반을 가른다.
- 옅은 소금물에 배추를 담갔다가 꺼낸다.
- 배추에 굵은소금을 골고루 뿌려 1시간 반 정도 절인다.
- 뒤집어 다시 1시간 반 정도 절인다.
- 적당히 절여지면 물로 잘 헹군 후 물기를 뺀다.

2. 무 색들이기
- 채 썬 무에 고춧가루 1/2컵을 넣고 버무려 색을 낸다.

3. 속 만들기
- 물에 찹쌀가루를 풀어 끓여 찹쌀풀을 만든다.
- 찹쌀풀에 고춧가루 1/2컵을 넣어 불린다.
- 양념을 모두 넣어 섞는다.
- 부재료를 모두 넣고 살짝 버무린다.

4. 속 넣기
- 절인 배추의 뒤쪽부터 버무린 김치 속을 고루 바른다.
- 배춧잎 사이사이에 소를 펴서 넣는다.
- 겉잎으로 전체를 싼다.

5. 담아 보관하기
- 배추 자른 단면을 위로 하여 항아리에 차곡차곡 담는다.

Ingredients

Main ingredients
Napa Cabbage 3.4kg · Sea Salt 1C

Sub-ingredients
Korean Radish 1.6kg (julienne) · Scallion 100g (cut into 4cm
length) · Korean watercress 150g (cut into 4cm length) · Sweet Rice
Powder 3T · Water 1½C

Seasonings
Red Chili Powder 1C (1/2C for radish, 1/2C for rice paste) · Fish Sauce
3T · Salted Shrimp 3T (finely chopped) · Sugar 2T · Garlic 4T
(minced) · Ginger 2t (minced) · Salt 2T

Steps

1. Salting cabbage
- Remove outer leaves of cabbages and halve them.
- Rinse with light brine.
- Salt over and into the cabbages using sea salt. Set aside for
  about 1½ hours.
- Turn over and set aside for another 1½ hours.
- If well-salted (softened) rinse well with water and drain.

2. Coloring radish
- Add 1/2 cup of chili powder into radish and toss.

3. Make stuffing
- Make rice paste by boiling water with rice powder.
- Add 1/2 cup of chili powder into cooled rice paste.
- Add all remaining seasonings.
- Mix the seasonings with colored radish, scallion and dropwort.

4. Filling/Stuffing cabbages
- Spread the mix (no. 3) from outer leaves of cabbage.
- Stuff the cabbage layer by layer.
- Wrap the filled cabbage up using the outer leaf.

5. Storing
- Stack them showing the cut side up into a jar.

203

귀띔 / Tips

1. 김치의 가장 적당한 보관온도는 0~5℃ 정도이다.
2. 절이는 소금의 양과 시간은 계절이나 상황에 따라 바뀐다.

1. The proper storage temperature of Kimchi is between 0~5℃.
2. The amount of salt and duration of salting can vary depending on various conditions.

보쌈

돼지고기를 푹 삶아 한입 크기로 썰어낸 수육을 절인 배추에 싸서 새우젓을 곁들여 먹는 음식이다.
김장하는 날 즐겨 먹는다.

Bossam_ *Boiled Pork Belly Served with Vegetables*

Bossam is boiled pork belly served with vegetables. The pork is sliced into bite-size pieces.
It is one type of 'wrap dishes' in Korean: it can be enjoyed by placing boiled pork, salted shrimp, and
seasoned vegetables on salted Korean cabbage.

재료

주재료
돼지고기 600g (삼겹살/ 목살) · 요리용 실 · 물 8C · 배추 1/2포기 (=800g) · 소금 35g · 물 3C

부재료
된장 1T · 소금 1t · 대파 1대 (5cm 길이) · 생강 1쪽 (납작썰기) · 통마늘 7쪽 · 통후추 10알 · 청주 1T

새우젓
새우젓 3T (굵게 다진 것) · 고춧가루 1/2T · 깨소금 1t · 다진 마늘 1t · 다진 파 2t

만드는 법

1. 배추잎 절이기
- 배추는 잎을 떼어 윗부분을 잘라버리고 소금물에 절인다.
- 헹궈서 물기를 뺀다.

2. 돼지고기 삶기
- 돼지고기는 실로 묶어준다.
- 냄비에 고기가 잠길 정도로 물을 붓는다.
- 된장을 체에 걸러 풀어 넣는다.
- 물이 끓기 시작하면 돼지고기와 대파, 생강, 마늘, 통후추, 청주, 소금을 넣고 삶는다.
- 1시간쯤 삶은 뒤 바로 건져내 식힌다.
- 한 김 식힌 고기를 썬다.

3. 새우젓 양념하기
- 다진 새우젓에 새우젓 국물과 양념을 넣고 고루 섞는다.

4. 그릇에 담기
- 삶은 돼지고기와 절인 배추를 접시에 담는다.
- 새우젓을 곁들여낸다.

Ingredients

Main ingredients
Pork 600g (belly or neck fillet, blooded) · Cooking Thread · Water 8C · Napa Cabbage 800g · Salt 35g · Water 3C

Sub-ingredients
Soy Bean Paste 1T · Salt 1t · Green Onion 1 stem (5cm length) · Ginger 30g (sliced) · Garlic 7 cloves · Black Pepper Whole 10ea · Rice Wine 1T

Salted shrimp dip sauce
Salted Shrimp 3T (chopped) · Red Chili Powder 1/2T · Sesame Seeds 1t (ground) · Garlic 1t (minced) · Green Onion 2t (finely chopped)

Steps

1. Salting cabbage
- Soak Korean cabbage leaves in salt water.
- Rinse with water and drain.

2. Boiling pork
- Tie pork with cooking thread.
- Add pork and water in a pot.
- Add soy bean paste using a sieve.
- When the water starts boil, add the pork and rest of sub-ingredients.
- Boil for an hour and take pork out. Let it cool for a while.
- Slice the boiled pork into 0.5cm thick pieces.

3. Making dip
- Mix all seasonings for salted shrimp dip.

4. Plating
- Serve sliced pork and salted cabbage in a plate.
- Serve with salted shrimp dip.

보쌈무생채

보쌈용 무생채는 채 썰어 소금에 절였다가 고춧가루와 양념을 넣고 무쳐 먹는 음식이다.

Bossam-mu-saengchae

_ Spicily Seasoned Radish for Bossam

Bossam-mu-saengchae is julienned white radish seasoned with red chili powder and other ingredients.

재료

주재료
무 400g (0.7×0.7cm로 굵은 채 썰기) · 소금 2T · 설탕 3T · 물엿 2T · 물 1/2C · 생굴 1/2C

부재료
대파 15g (흰 부분 어슷썰기) · 쪽파 20g (4cm로 썰기) · 미나리 40g (4cm로 썰기) · 생률 3개 (납작썰기)

양념
굵은 고춧가루 4T · 다진 마늘 1t · 다진 생강 1/2t · 멸치액젓 1T · 설탕 1t · 잣 1T

고명
통깨 1t

만드는 법

1. 무 절이기
- 채 썬 무에 소금, 설탕과 물엿을 넣고 절인다.

2. 양념 만들기
- 액젓과 갖은 양념을 고루 섞는다.

3. 무생채 버무리기
- 양념에 무채를 넣어 버무리고 부재료를 섞는다.

4. 그릇에 담기 및 고명 올리기
- 무생채와 생굴을 나란히 담고 고명으로 통깨를 뿌린다.

Ingredients

Main ingredients
Radish 400g (julienne (0.7×0.7cm)) · Salt 2T · Sugar 3T · Corn Syrup 2T · Water 1/2C · Fresh Oyster 1/2C

Sub-ingredients
Green Onion 15g (bias-cut, white portion) · Scallion 20g (4cm, length) · Korean Watercress 40g (4cm, length) · Chestnuts 3ea (sliced)

Seasonings
Red Chilli Powder 4T · Garlic 1t (minced) · Ginger 1/2t (minced) · Fish Sauce 1T · Sugar 1t · Pine Nuts 1T

Garnish
Sesame Seeds 1t

Steps

1. Salting radish
- Add salt, sugar and corn syrup to radish.

2. Make seasoning

3. Tossing
- Gently toss the radish with seasoning and add the sub-ingredient.

4. Plating
- Transfer to a plate along with fresh oyster and sesame seeds on top.

우거지된장국

우거지는 배추를 다듬을 때, 골라놓은 겉잎을 말한다.
예로부터 한국에서는 김장을 담그는 날이면 다듬고 남은 우거지를 말려서 저장하였다가 국을
끓였다.

Ugeoji-doenjang-guk

_ *Korean Cabbage Soup with Soy Bean Paste*

Ugeoji refers to the trimmed-off leaves of leafy vegetables such as cabbages. In fall, Koreans traditionally
prepare Kimchi for the winter. While making Kimchi, they gather Ugeoji from cabbages and enjoy a
soup with Ugeoji and soy bean paste.

재료

주재료
배추잎 250g

부재료
마른 멸치 15g · 대파 20g (어슷썰기) · 물 8C

양념
된장 4T · 다진 마늘 1t · 소금 약간

만드는 법

1. 배추 손질
- 배추잎을 한 잎씩 떼어낸다.
- 끓는 물에 소금을 넣고 데쳐낸다.
- 찬물에 헹구어 꼭 짜둔다.

2. 육수 만들기
- 다듬은 멸치를 팬에 넣고 물기 없이 볶는다.
- 물을 넣고 끓인다.
- 멸치를 건져낸다.
- 체를 이용해 된장을 푼다.

3. 우거지 넣고 끓이기
- 데친 배추잎을 1.5cm 길이로 썬다.
- 육수가 끓으면 배추를 넣고 무르도록 끓인다.

4. 간하기
- 다시 끓으면, 마늘과 소금으로 간한다.
- 대파를 넣고 불을 끈다.
- 국그릇에 담아낸다.

Ingredients

Main ingredients
Napa Cabbage, leaves 250g

Sub-ingredients
Dried Anchovy 15g · Green Onion 20g (bias-cut) · Water 8C

Seasonings
Soy Bean Paste 4T · Garlic 1t (minced) · Salt pinch

Steps

1. Proparing cabbage
- Tear cabbage leaf by leaf.
- Blanch them in boiling water with salt.
- Squeeze off excess water and set aside.

2. Making anchovy broth with soy bean paste
- Toast trimmed anchovy in a pan.
- Add water and bring it to boil.
- Take anchovies out of the broth.
- Add soy bean paste using a sieve to dissolve.

3. Adding blanched napa cabbage leaves
- Cut blanched cabbage into 1.5cm length.
- When the broth starts to boil and add the Ugeoji.

4. Seasoning
- When it starts boiling again, season with minced garlic and salt.
- Add green onion and turn off the heat.
- Transfer to a soup bowl.

귀띔
Tips
1. 풋내를 없애기 위해 배추를 데쳐낸다.
2. 멸치를 볶아야 국물 맛이 비리지 않다.

1. Blanching the cabbage leaves prevents the soup from having a grassy odor.
2. Toasting the anchovies prevents the off—fishy odor of the broth.

김치만둣국, 고기버섯산적, 장김치

온 가족이 모여 이야기를 나누며 함께 빚는 김치만두는 가족의 정을 두텁게 하고, 저마다의
솜씨가 묻어난 만두를 먹을 때 정겨움이 느껴진다.

Kimchi-mandu-guk, Gogi-beoseot-sanjeok, Jang-kimchi

When a family makes Kimchi-mandu together, they build strong bonds through communicating.
People have fun and feel affection when eating Mandu with different characteristics.

김치만둣국

추운 겨울 인기 식단으로 김치를 넣은 만둣국을 꼽을 수 있다. 특히 평안도나 황해도, 강원도에는 명절에 떡국 대신 만둣국을 해 먹는 경우가 많다.

Kimchi-mandu-guk

_ Kimchi Dumpling Soup

Kimchi-mandu-guk is one of the most beloved winter dishes in Korea. People from the north provinces enjoy this dish on New Year's Day instead of tteok-guk (Rice cake soup).

재료

주재료
김치 100g (속을 털어낸 뒤 송송 썰기) • 마른 표고 2장 (불린 후 채 썰기) • 두부 50g (물기 제거 후 으깨기) • 숙주 100g (데친 후 송송 썰기) • 대파 1대 (송송 썰기) • 다진 쇠고기 100g

쇠고기양념
소금 2/3t • 다진 파 2t • 다진 마늘 1t • 깨소금 · 참기름 각 1t씩 • 후추 약간

만두소양념
다진 파 · 다진 마늘 · 깨소금 · 참기름 각 1t씩

부재료
만두피 재료 • 밀가루 2C • 소금 1t • 물 6~7T (따뜻한 물)

만두육수
양지머리 200g (핏물 제거) • 대파 1/2대 • 통마늘 3톨 • 통후추 약간 • 물 6C

고명
달걀지단 27쪽 (지단 만들기 참조할 것)

만드는 법

1. 만두피 반죽
- 밀가루에 소금과 물을 넣고 반죽한 뒤 젖은 면보로 싸둔다.
- 밀가루 반죽을 얇게 밀어 지름 7cm 되는 둥근 틀로 찍어준다.

2. 만두육수와 편육 만들기
- 끓는 물에 육수재료를 넣고 고기가 익을 때까지 삶아준다.
- 다 익은 고기는 젖은 면보에 싸 눌러 편육으로 썬다.
- 육수는 기름을 걷어낸다. (체 혹은 면보 사용)

3. 만두소 만들기
- 주재료에 만두소양념을 넣고 고루 섞이도록 반죽한다.

4. 만두 빚기
- 만두피에 만두소 1T씩 넣고 반달형으로 접어 끝에 물을 묻혀 붙인 뒤 양쪽 끝을 다시 붙게 한다.

5. 끓이기
- 2의 육수가 끓으면 만두를 넣고 다진 마늘(1t)과 소금, 국간장으로 간을 한다. 만두가 떠오르면 불을 끈다.

6. 그릇에 담아내기
- 그릇에 만두국을 담고 편육과 지단으로 고명을 올려낸다.

Ingredients

Main ingredients
Kimchi 100g (stuffing removed, chopped) • Dried Pyogo Mushroom 2ea (soaked, julienne) • Dubu 50g (water squeezed, mashed) • Mung Bean Sprout 100g (blanched, chopped) • Green Onion 1 stem (chopped) • Beef 100g (minced)

Seasonings for beef
Salt 2/3t • Green Onion 2t (finely chopped) • Garlic 1t (minced) • Sesame Seeds1t (ground) • Sesame Oil 1t • Black Pepper pinch (ground)

Seasonings for stuffing
Salt 1t • Green Onion (finely chopped), Garlic (minced), Sesame Seeds (ground), Sesame Oil 1t/ each

Sub-ingredients
Dumpling skin • Flour 2C • Salt 1t • Water 6~7T (warm)

Broth/Stock
Beef, Brisket 200g (blooded) • Green Onion 1/2 stem • Garlic 3 cloves • Black Pepper pinch (whole) • Water 6C

Garnish
Egg Jidan (refer to page 27~28)

Steps

1. Making dumpling skin
- Mix all sub—ingredients to make dough.
- Set aside wrapping with a wet cloth.
- Roll up dough and press 7cm diameter skins.

2. Making broth and boiled beef garnish
- Cook brisket in boiling water.
- Cool the beef down pressing with a heavy object.
- Slice the cooked beef to use for garnish.
- Skim fat off the broth.

3. Making dumpling stuffing
- Mix all the main ingredients and seasonings.

4. Shaping
- Put a spoonful of stuffing on a skin and wet the edge.
- Fold up to shape a half moon.
- Connect both ends to shape a beautiful round.

5. Boiling
- Bring the broth (no. 2) to a boil and add shaped dumplings.
- Season with minced garlic, salt and gukganjang.
- When dumplings float up, turn off heat.

6. Plating
- Transfer to a soup bowl and garnish with boiled beef and Jidan.

213

고기버섯산적

고기버섯산적이란 고기와 버섯을 꼬치에 번갈아 꿰어 석쇠에 구워낸 음식이다.

Gogi-beoseot-sanjeok

__ *Beef and Mushroom Skewer*

Gogi-beoseot-sanjeok is one of the representative meat skewers in Korean cuisine. Sliced meat and mushroom are skewered alternately and grilled.

재료

주재료
새송이버섯 2~3개 (0.7×7cm 썰기) • 쇠고기 200g (우둔살, 1×8×0.7cm로 썰기) • 쪽파 8줄기 (7cm 길이로 썰기)

쇠고기양념
간장 2T • 설탕 3/4T • 다진 파 1T • 다진 마늘 1/2T • 깨소금 1/2T • 참기름 1/2T • 후춧가루 약간

버섯 유장
소금 1t • 참기름 1T

고명
잣가루 1/2t

만드는 법

1. 양념하기
- 쇠고기는 칼집을 넣어 양념에 재운다.
- 새송이와 쪽파는 유장을 발라준다.

2. 꼬치에 끼우기
- 새송이-고기-실파-고기-새송이 순으로 꼬치에 끼운다.

3. 굽기
- 2를 프라이팬이나 석쇠에 구워낸다.

4. 그릇에 담기
- 접시에 3의 산적을 담고 잣가루를 뿌려낸다.

Main ingredients
Saesongi Mushroom 2~3ea (0.7×7cm sliced) • Beef 200g (fore rump, 1×8×0.7cm thick strips) • Scallion 8ea (7cm length)

Seasonings for beef
Soy Sauce 2T • Sugar 3/4T • Green Onion 1T (finely chopped) • Garlic 1/2T (minced) • Sesame Seeds 1/2T (ground) • Sesame Oil 1/2T • Black Pepper pinch (ground)

Oil seasonings (yujang)
Salt 1t • Sesame Oil 1T

Garnish
Pine Nuts, ground 1/2t

Steps

1. Seasoning
- Give small slits on beef and marinate with seasonings.
- Brush or rub oil seasoning on mushrooms and scallions.

2. Skewering
- Skewer ingredients in order. (mushroom-beef-scallion-beef-mushroom)

3. Cooking
- Pan-fry or grill the skewers.

4. Plating
- Transfer to a plate and garnish with ground pine nuts.

215

장김치

장김치는 간장에 절인 배추와 무를 여러 재료와 섞어 젓갈이나 고춧가루가 아닌 간장으로 간을
맞추어 담근 김치이다. 조선시대 궁중이나 대갓집에서 설날이나 추석에 만들어 먹던 김치로 대추,
밤, 배, 표고버섯, 잣 등 김치재료가 호화로워 서민적이지는 않으나 정월 떡국상이나 잔칫상 등에
올렸다. 잘 익은 장김치는 간장의 색과 향이 조화를 이루어 별미다.

Jang-kimchi

__ *Watery Kimchi Made with Soy Sauce*

Jang-kimchi has a very unique feature: the base is made with soy sauce, which is quite different from
other kinds of Kimchi. With its expensive ingredients, jang-kimchi was enjoyed by only the rich in old
days.

재료

주재료
배추속대 500g (2.5×3cm, 썰기) · 무 500g (2.5×3×0.5cm, 썰기) · 간장 1½C

부재료
밤 8개 (4개는 채 썰기, 4개 편썰기) · 배 300g (무와 같은 크기) · 건표고버섯 3장 (불린 뒤 채 썰기) · 갓 50g (3cm 길이로 썰기) · 미나리 50g (3cm 길이로 썰기) · 대파 흰 부분 30g (3cm 길이로 채 썰기) · 대추 6개 (씨 제거 후 채 썰기) · 석이버섯 4장 (불린 후 채 썰기) · 마늘 20g (가늘게 채 썰기) · 생강 10g (가늘게 채 썰기) · 실고추 약간 (3cm 자르기) · 잣 1t · 물 10C (끓인 후 식혀 사용) · 설탕 1T

만드는 법

1. 간장에 절이기
- 먼저 배추속대에 간장 1 cup을 넣고 색을 들인다.
- 무와 나머지 간장 1/2 cup을 넣고 고루 버무려 절인다.
- 절여지면 간장을 따라낸다.

2. 모든 재료 섞기
- 1번과 나머지 부재료를 골고루 섞어준다.

3. 장김칫국 만들기
- 따라낸 간장에 물 10 cup과 설탕 1T을 넣어준다.

4. 숙성시키기
- 2와 3을 섞는다.
- 잣과 실고추를 넣고 냉장고에서 숙성시킨다.

5. 그릇에 담아내기

Ingredients

Main ingredients
Napa Cabbage 500g (soft inner leaves only, 2.5×3cm sliced) · Radish 500g (2.5×3×0.5cm sliced) · Soy Sauce 1½C

Sub-ingredients
Chestnuts 8ea (4 for fine julienne, 4 for slices) · Korean Pear 300g (same size with radish) · Dried Pyogo Mushroom 3ea (soaked, julienne) · Mustard Leaves 50g (3cm length) · Korean Watercress 50g (3cm length) · Green Onion 30g (white part only, 3cm julienne) · Jujube 6ea (seeded, julienne) · Stone Ear Mushroom 4ea (soaked, julienne) · Garlic 20g (fine julienne) · Ginger 10g (fine julienne) · Silgochu pinch (3cm length) · Pine Nuts 1t · Water 10C (boiled and chilled) · Sugar 1T

Steps

1. Coloring/ Marinating
- Mix cabbage and soy sauce (1C) to color. Set aside.
- Add radish and remaining soy sauce (1/2 cup). Set aside.
- Minutes later, strain soy sauce.

2. Mixing
- Mix colored cabbage and radish with sub-ingredients.

3. Making watery kimchi base
- Mix strained soy sauce with water and sugar.

4. Storing for fermentation
- Mix no. 2 and 3 in a container.
- Add silgochu and pine nuts.
- Store in a refrigerator to ferment.

5. Plating

외국인도 빠져드는 한국밥상

귀띔
Tips

1. 기온에 따라 배추 절이는 시간과 숙성 시간은 다를 수 있다.
2. 장김치는 주안상이나 떡국, 교자상에 잘 어울리는 음식으로 차게 낸다.

1. The duration of marinating could vary, depending on weather conditions or temperature.
2. Jang-kimchi goes well with dumpling soup or rice-cake-soup (Tteok-guk). Serve cold.

구절판 *Gujeolpan* / 골동면(비빔국수) *Goldong-myeon(Bibim-guksu)* / 너비아니구이 *Neobiani-gui*

/ 연근더덕무침 *Yeongeun-deodeok-muchim* / 채소간장장아찌 *Chaeso-ganjang-jangajji* / 장조림 *Jang-jorim*

/ 탕평채 *Tangpyeong-chae* / 팥죽 *Pat-juk* / 오이소박이 *Oi-sobagi* / 무생채 *Mu-saengchae*

별
식

주안상

주안상이란 술과 안주를 차려 놓은 상을 의미한다. 예로부터 한국의 각
가정에서는 계절별로 각종 술을 빚어 놓아, 귀한 손님이 오셨을 때 좋은 술과
그에 걸맞은 안주를 대접하는 것이 상례였다.

Juansang

Juansang refers to a table with drink and accompaniments on top. korea has
a tradition, where each household creates various seasonal drinks, and when
important guests pay a visit to one's house, they were treated with a high-quality
drink and some snacks that blend well with them.

구절판

구절판은 아홉 칸으로 나누어진 그릇 이름인데, 그대로 음식 이름이 된 것이다. 아홉 칸으로 나뉜 그릇의 가운데 밀전병을 담고 가장자리 칸에는 쇠고기, 채소, 달걀지단, 버섯 등을 담아서 골고루 싸서 먹는 음식이다. 색이 화려하고 맛이 산뜻해 전채음식으로 잘 어울린다.

Gujeolpan

__ Korean Traditional Wrap Dish Served on a Nine-sectioned Plate

The name 'Gujeolpan' actually came from the shape of its plate, 'nine-sectioned-plate'. Miljeonbyeong, flour crepe, is placed at the center and other ingredients surround it. Gujeolpan is eaten after other ingredients are wrapped in Miljeonbyeong. Since it is tasty, refreshing and colorful, Gujeolpan is good as an appetizer.

재료

주재료
쇠고기 120g (얇게 채 썰기) · 건표고버섯 5장 (불려서 채 썰기) · 석이버섯 3장 (불려서 채 썰기) · 오이 1개 (채 썰기, 소금과 기름 필요) · 당근 1/3개 (채 썰기, 소금과 기름 필요) · 숙주 100g (머리제거) · 달걀 3개

부재료
밀전병 · 밀가루 1C · 물 1¼C · 소금 1/2t
겨자초장 · 겨잣가루/연겨자 2T · 물 1T · 식초 1T · 설탕 1T · 간장 약간
고기양념 · 간장 2T · 설탕 2T · 다진 파 2T · 다진 마늘 1T · 참기름 1/2T · 깨소금 약간 · 후추 약간

만드는 법

1. 쇠고기와 버섯 양념해 볶기
- 쇠고기와 버섯에 양념을 반씩 넣어 볶는다.
2. 오이, 당근, 숙주, 석이버섯 준비하기
- 오이는 소금에 살짝 절여 꼭 짠 후, 팬에 기름을 두르고 볶아낸다.
- 당근은 팬에 기름을 두르고 소금 간을 해서 볶는다.
- 숙주는 데친 후, 소금과 참기름으로 무친다.
- 석이버섯은 팬에 기름을 두르고 소금 간을 해서 볶는다.
3. 달걀지단 만들기
- 달걀은 황 · 백으로 분리하여 지단을 부친다.
- 4cm 길이로 곱게 채 썬다.
4. 밀전병 부치기 (7~8cm 지름)
- 재료를 섞은 후, 한 숟가락 분량씩 동그랗게 부친다.
5. 구절판에 담기
- 가운데 밀전병, 나머지 칸에 다른 재료를 넣는다.
- 겨자초장과 함께 낸다.

Ingredients

Main ingredients
Beef 120g (thinly julienne) · Dried Pyogo Mushroom 5ea (soaked, julienne) · Stone Ear Mushroom 3ea (soaked, julienne) · Cucumber 1ea (julienne) · Carrot 1/3ea (julienne) · Mung Bean Sprout 100g (head off) · Egg 3ea

Sub-ingredients
Miljeonbyeong (flour crepe) · Flour 1C · Water 1¼C · Salt 1/2t
Vinegar Mustard Dip · Korean Mustard Paste (or Powder) 2T · Water 1T · Vinegar 1T · Sugar 1T · Soy Sauce to taste
Seasonings · Soy Sauce 2T · Sugar 2T · Green Onion 2T (finely chopped) · Garlic 1T (minced) · Sesame Oil 1/2T · Sesame Seeds pinch (ground) · Black Pepper pinch

Steps

1. Seasoning and stir-frying beef and dried pyogo mushroom
- Mix a half of seasonings to beef and another half to mushroom. Stir-fry them separately in a pan with oil.
2. Cooking cucumber, carrot, sprout & ear mushroom
- Salt cucumber for few minutes to soften. Squeeze off excess water and stir-fry in a pan with oil.
- Stir-fry julienned carrot using cooking oil and salt.
- Blanch mung bean sprouts and season with salt and sesame oil.
- Stir-fry stone ear mushroom using cooking oil and salt.
3. Making egg jidan
- Separate yolk and white. Beat lightly adding salt.
- Pan-fry over low heat with little oil.
- Julienne them into 4cm length.
4. Making miljeonbyeong (7~8cm diameter)
- Mix ingredients in a bowl.
- Pan-fry a spoonful over low heat with little oil.
5. Plating and serving
- Put miljeonbyeong in the middle. Surround other ingredients.
- Serve with gyeojachojang, vinegar mustard dip.

223

귀띔 / Tips

1. 밀전병을 부드럽게 부치려면 반죽을 체에 한 번 내린다.
2. 밀전병 사이에 잣가루를 뿌리면 서로 붙지 않고 맛도 좋다.
3. 상차림할 때 먼저 내놓는 전채음식으로 좋고, 모양과 색이 화려해 좋은 분위기가 된다.

1. Sieve Miljeonbyeong mix to aid softer texture.
2. Sprinkle pine nuts powder to separate miljeonbyeong easily and to taste better.
3. Gujeolpan is a representative appetizer in Korean cuisine and gives good color to table setting.

골동면 (비빔국수)

골동면은 삶은 소면에 볶은 고기와 표고버섯, 오이 등 여러 가지 재료를 간장양념으로 버무린 국수로 비빔국수라고도 한다.

Goldong-myeon (Bibim-guksu)

__ *Noodles Mixed with Various Ingredients*

Goldong-myeon is a traditional noodle dish made with Somyeon. Somyeon is seasoned with soy sauce and other various ingredients such as beef, mushroom, cucumber, etc.

재료

주재료
소면 300g

부재료
소고기 100g (우둔살, 굵게 다지기) · 건표고버섯 3장 (불린 후 채 썰기) · 오이 1개 (오이를 길이로 길게 잘라 씨부분을 도려내고 어슷썰기) · 달걀 2개 (황 · 백 지단 만들기, 5cm 길이로 채 썰기)

고기 · 표고버섯 양념
간장 1½T · 설탕 2t · 다진 파 2t · 다진 마늘 1t · 깨소금 1t · 참기름 1t · 후추 약간

소면양념
간장 3T · 설탕 1T · 깨소금 1T · 참기름 1T

고명
실고추 약간

만드는 법

1. 재료 준비
- 다진 소고기와 채 썬 표고버섯을 각각 양념한다.
- 어슷썰기한 오이는 소금물에 절여 물기를 완전히 제거한 뒤 달군 팬에 기름을 넣고 재빨리 볶아 식힌다.
- 달군 팬에 기름을 넣고 소고기를 볶다가 물 3큰술을 넣고 살짝 끓인 뒤 식힌다.
- 달군 팬에 기름을 약간 두르고 표고버섯을 볶아 식힌다.

2. 국수 삶기
- 끓는 물에 국수를 넣고 끓어오르면 물 1C을 두세 번에 나눠 넣는다.
- 하얀 심이 없어질 때까지 삶아 찬물에 재빨리 넣어 손으로 살살 비벼가며 여러 번 헹궈준다.
- 면을 체에 넣고 물기를 제거해 준다.

3. 밑간하기
- 달걀지단을 제외한 오이, 고기, 표고버섯을 국수와 함께 섞고 소면양념을 넣어 살살 버무린다.

4. 담아내기
- 그릇에 담고 고명으로 지단과 실고추를 올려낸다.

Ingredients

Main ingredients
Somyeon 300g

Sub-ingredients
Beef 100g (Rump, knife-minced) · Dried Pyogo Mushroom 3ea (soaked, julienne) · Cucumber 1ea (halved and bias-cut) · Egg 2ea (separated yolk and white and pan-fried, julienne in 5cm length)

Seasonings for beef & mushroom
Soy Sauce 1½T · Sugar 2t · Green Onion 2t (finely chopped) · Garlic 1t (minced) · Sesame Seeds 1t (ground) · Sesame Oil 1t · Black Pepper pinch

Seasonings for noodle
Soy Sauce 3T · Sugar 1T · Sesame Seeds 1T (ground) · Sesame Oil 1T

Garnish
Silgochu pinch

Steps

1. Preparing sub-ingredients
- Marinade beef and mushroom with seasonings.
- Soak cucumber in brine for minutes and squeeze—drain. Stir—fry in a pre—heated pan quickly with cooking oil and cool down.
- Stir—fry beef in a pre—heated pan quickly with cooking oil and add 3 spoons of water. Let evaporate and cool down.
- Stir—fry mushroom in a pre—heated pan quickly with cooking oil and cool down.

2. Boiling noodle
- Put noodles in a pot of boiling water and add a cup of cold water 2~3 times when starting to boil.
- When the noodle are cooked completly, transfer to a pot of ice cold water and rub.
- Transfer to a strainer to remove excess water.

3. Mixing
- Add all prepared ingredients and seasonings for noodles except eggs.
- Mix gently.

4. Plating
- Transfer to a bowl and garnish with eggs and chili.

귀띔 Tips

1. 간장 대신 고추장양념을 해도 좋으며 면은 먹기 직전에 삶아야 좋다.
2. 달걀지단 만드는 법은 잡채 조리법을 참조한다. (103쪽)

1. Boil noodles right before serving to prevent from being over—cooked.
2. Making egg jidan refers to the Japchae recipe. (page 103)

너비아니구이

너비아니란 고기를 도톰하게 저며 양념하여 불에 직접 굽는 음식이다.

Neobiani-gui_ *Pan-fried or Grilled Beef*

Neobiani-gui is thickly sliced (but not too thick) grilled beef with soy sauce seasoning.

재료

주재료
쇠고기 500g (등심 0.5~0.7cm 두께 썰기)

부재료
계절 채소 (양파 50g, 양배추 100g, 영양부추 100g) • 샐러드 소
스 (간장 1T, 설탕 1/2T, 식초 1T, 참기름 1T)

쇠고기양념
간장 4T • 배즙 4T • 설탕 2T • 다진 파 3T • 다진 마늘
1½T • 깨소금 1½T • 참기름 1½T • 후추 약간

고명
잣가루 약간

만드는 법

1. 쇠고기 준비
  -쇠고기는 잔칼집을 넣어 연하게 만든다.
2. 양념 만들기
  -그릇에 양념을 모두 넣고 섞는다.
3. 고기 재우기
  -양념에 고기를 넣고 주무른다.
  -30분간 재워둔다.
4. 굽기
  -예열된 팬에 고기를 굽는다.
5. 계절 채소 샐러드 만들기
  -채썬 채소를 소스에 무쳐낸다.
6. 담아내기
  -그릇에 완성된 너비아니를 옮겨담고 잣가루를 뿌려낸다.

Ingredients

Main ingredients
Beef 500g (Sirloin, 0.5~0.7cm thick)

Sub-ingredients
Seasonable Vegetables (Onion 50g, Cabbage 100g, Buchu
100g) • Dressing (Soy Sauce 1T, Sugar 1/2T, Vinegar 1T, Sesame Oil 1T)

Seasonings for beef
Soy Sauce 4T • Pear Juice 4T • Sugar 2T • Green Onion
3T (minced) • Garlic 1½T (minced) • Sesame seeds 1½T
(ground) • Sesame Oil 1½T • Black Pepper pinch

Garnish
Pine Nuts pinch (ground)

Steps

1. Preparing beef
  -Give slits in the beef for tenderizing.
2. Making seasoning
  -Mix all seasonings in a big bowl.
3. Marinating
  -Add no. 1 into no. 2 and mix well.
  -Marinating for 30 minutes.
4. Pan frying / grilling
  -Pan fry the marinated beef.
5. Making a salad
  -Mix seasonable vegetables with dressing.
6. Plating
  -Transfer to plate and garnish with pine nuts powder.

연근더덕무침

연근은 연꽃의 뿌리이다. 가을에서 겨울에 나며 아삭거리는 식감으로 주로 조림이나 무침으로 이용된다.

더덕 또한 땅속에서 자라는 뿌리채소로서 특유의 쌉싸래한 맛과 향이 좋다.

연근더덕무침은 연근의 식감과 더덕의 향이 새콤한 소스와 어우러져 입맛을 돋우는 음식이다.

Yeongeun-deodeok-muchim

__ Lotus Root and Mountain Herb Instant Pickle

Yeongeun means lotus root, which is harvested in fall and winter. Lotus root is crispy and often boiled down in soy sauce or other seasonings. Deodeok is the root of a mountain herb with an astringent taste and scent. Yeongeun-deodeok-muchim is a seasoned mix of Yeongeun and Deodeok, which satisfies one's appetite with its crunchy and slightly bitter taste. It is a mouth-watering summer food.

재료

주재료
연근 200g (껍질 벗겨서 0.3cm 둥글게 썰기) • 더덕 100g (껍질 벗겨 반 갈라 두들겨 결대로 찢기)

부재료
밤 4개 (도톰하게 편썰기) • 홍고추 1/4개 • 풋고추 1/4개 (얇게 어슷썰기) • 잣가루 1T • 통깨 1T

단촛물
설탕 2T • 식초 2T • 소금 2t • 레몬즙 1T • 물 6T

고명
잣가루 1T • 통깨 1T

만드는 법

1. 연근 데치기
-연근을 끓는 물 5컵에 식초 1큰술을 넣고 5분간 데친다.
2. 재우기
-단촛물을 만들어 1/2은 연근, 나머지는 더덕을 담가 30분 이상 차게 둔다.
3. 버무리기
-차게 두었던 연근과 더덕을 섞어서 준비된 밤, 홍고추, 풋고추, 잣가루, 통깨를 넣고 가볍게 무쳐준다.
4. 담아내기
-고명으로 잣가루와 통깨를 올린다.

Ingredients

Main ingredients
Lotus Roots 200g (peeled, sliced 0.3cm thick) • Deodeok 100g (peeled, cut in half, ponded, shredded)

Sub-ingredients
Chestnuts 4ea (thick sliced) • Red Chili 1/4ea (thinly bias-cut) • Green Chili 1/4ea (thinly bias-cut) • Pine Nuts 1T (ground) • Sesame Seeds 1T

Seasonings
Sugar 2T • Vinegar 2T • Salt 2t • Lemon Juice 1T • Water 6T

Garnish
Pine Nuts 1T (ground) • Sesame Seeds 1T

Steps

1. Blanching lotus roots
-Bring 5cups of water to a boil and add 1T of vinegar. Blanch it for 5 minutes.
2. Marinating
-Make vinegar water mix with all ingredients.
-Using 1/2 of vinegar water marinating lotus roots and the rest for marinating deodeok for 30 minutes.
3. Tossing
-In a bowl, mix all ingredients and toss well.
4. Plating
-Garnish with ground pine nuts and sesame seeds.

귀띔 연근과 더덕은 따로 차갑게 보관하였다가 필요할 때 섞어 사용한다.
Tips Prepare lotus root and deodeok separately in a refrigerator and combine when you need it.

채소간장장아찌

오이, 무, 양파 등의 각종 채소를 한입 크기로 썰어 초간장에 절여두고 먹는 장아찌이다.

Chaeso-ganjang-jangajji

_ Vegetables & Fruit Soy Sauce Pickle

Chaeso-ganjang-jangajji is pickled vegetables.
Various vegetables such as cucumber, white radish, and onion are sliced into bite-size pieces, and
pickled with soy sauce and vinegar.

재료

주재료

오이 3개 (0.5cm 두께로 둥글게 썰기) • 알타리무 3개 (0.5cm 두께로 둥글게 썰기) • 오이고추 3개 (0.5cm 두께로 둥글게 썰기) • 양파 1개 (한입 크기로 썰기) • 청양고추 1개 (0.5cm 두께로 둥글게 썰기) • 브로콜리 120g (한입 크기로 자르기)

배합초

간장 1C • 설탕 1½C • 통후추 1t • 월계수잎 2장 • 식초 1½C

만드는 법

1. 유리병에 채소를 담아준다.
2. 식초를 제외한 배합초를 끓인다.
3. 채소가 담긴 통에 식초를 먼저 붓고 끓는 배합초를 붓는다.
4. 이틀 후에 국물만 따라낸다.
5. 따라낸 국물을 끓인 후, 완전히 식힌다.
6. 식힌 국물을 다시 채소가 담긴 통에 붓는다.
7. 위 과정의 4~6번을 2~3회 반복한다.

Ingredients

Main ingredients

Cucumber/Young Radish/Cucumber Pepper, 3ea (0.5cm, slice)/each • Onion 1ea (bite-size)
Hot Geen Chili 1ea (0.5cm, slice) • Broccoli 120g (bite-size)

Pickling water

Soy Sauce 1C • Sugar 1½C • Whole Black Pepper 1t • Bay Leaves 2ea • Vinegar 1½C

Steps

1. Put all ingredients into a jar.
2. Boil all the pickling water mixture except vinegar.
3. Put the vinegar into the jar and add boiling pickling water.
4. After 2 days, strain pickling water in a pot.
5. Reboil the pickling water and let it cool down.
6. Put pickling water back to the jar.
7. Repeat the steps of no. 4 to no. 6 for 2~3 times.

231

장조림

쇠고기를 덩어리째 조림장을 넣고 질기지 않게 뭉근히 조려 두고두고 먹는 밑반찬이다.
채소를 곁들여 샐러드로 만들 수도 있고 꽈리고추 대신 달걀을 삶아 곁들여낼 수도 있다.

Jang-jorim_ *Beef Chunks Simmered in Soy Sauce*

Jang-jorim is beef chunks simmered and seasoned in soy sauce, which taste tender and not salty. Often Kkwari gochu (chili) or boiled eggs are simmered together with beef. It is a very popular traditional side dish in Korean meals. Sometimes it is mixed with vegetables and served as a salad.

재료

주재료
쇠고기 600g (홍두깨) · 물 15C (육수용)

부재료
향신채 · 마늘 5톨 (편썰기) · 생강 5g (편썰기)
통후추 10개 · 대파 1개 · 마른 고추 2개 · 꽈리고추
150g · 소금 약간

양념
조림장 · 간장 1C · 설탕 5T · 맛술 1T
육수 2C

만드는 법

1. 쇠고기 핏물 빼기
- 끓는 물에 쇠고기를 한 번 데쳐낸다.

2. 육수 만들기
- 찬물 15컵에 쇠고기와 향신채소를 넣고 50분 정도 끓인다.
- 쇠고기는 건져내고 육수는 체에 걸러낸다.

3. 꽈리고추 손질하기
- 꼬치를 이용하여 꽈리고추에 구멍을 낸다.
- 소금물에 살짝 데친다.

4. 조리기
- 냄비에 삶은 쇠고기와 육수, 조림장을 넣고 중불에서 졸인다.
- 국물이 반으로 졸아들면 꽈리고추를 넣고 끓인다.

5. 담아내기
- 그릇에 결대로 찢어서 꽈리고추와 함께 담아낸다.

Ingredients

Main ingredients
Beef 600g (chunk) · Water 15C

Sub-ingredients
Aromatic Ingredients · Garlic 5ea (sliced) · Ginger 5g
(sliced) · Black Pepper whole 10ea · Green Onion 1ea · Dried Red
Chili 2ea · Kkwari Gochu 150g · Salt pinch

Seasonings
Sauce · Soy Sauce 1C · Sugar 5T · Cooking Wine 1T · Broth 2C

Steps

1. Blanching beef
- Blanch beef in boiling water.

2. Making broth
- Add blanched the beef, aromatic vegetables and 15 cups of water in a pot.
- Boil for 50 min.
- Take the beef out & drain the beef broth with a fine sieve.

3. Preparing Kkwari Gochu
- Prick holes on the Kkwari Gochu several times with a skewer.
- Blanch Kkwari Gochu in salted boiling water.

4. Simmering
- In a pot, add the broth, beef, seasoning sauce and simmer on medium heat.
- When the broth is reduced in half, add Kkwari Gochu and boil for min.

5. Plating
- Tear beef into bite size and serve.

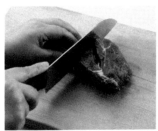

와인에도 빠져드는 한끼밥상

탕평채

녹두녹말로 만든 청포묵에 채소와 볶은 쇠고기를 넣어 새콤달콤한 양념으로 버무려낸 무침이다.

Tangpyeong-chae

__ *Mung Bean Jelly Mixed with Various Vegetable and Meat*

Tangpyeong-chae is a mix of mung bean jelly, stir-fried beef, and various vegetables.
It is seasoned with sweet and sour sauce.

재료

주재료
청포묵 300g (4×0.5×0.5cm, 채 썰기)

부재료
쇠고기 50g (우둔, 채 썰기) · 미나리 70g (줄기만, 4cm 길이로 자르기) · 숙주 70g (머리와 꼬리 다듬기)

양념
묵양념 · 참기름 1/2t · 소금 1/4T
쇠고기양념 · 간장 1/2T · 참기름 1/2t · 설탕 1/2t · 다진 마늘 1/2t · 다진 파 1t · 통깨 1/2t · 후추 약간
초간장 · 간장 1T · 설탕 1T · 식초 1T · 물 1T

고명
달걀 지단(4×0.2cm 썰기) 1장 · 홍고추 1/2개(2cm 썰기) · 김 (구워 잘게 부순 것) 1/2장

만드는 법

1. 묵 준비
- 끓는 물에 묵을 데친 후, 건져내어 물기를 뺀다.
- 참기름을 넣어 달라붙지 않게 한다.
- 소금으로 간을 한다.

2. 쇠고기 양념해 볶기
- 10분간 양념에 재워둔다.
- 볶아낸다.

3. 채소 데치기
- 미나리와 숙주는 끓는 물에 소금을 넣고 데쳐낸다.
- 찬물에 씻어내어 꼭 짠다.
- 미나리는 4cm 길이로 자른다.

4. 초간장 만들기
- 분량의 재료를 섞어 초간장을 만든다.

5. 버무리기
- 묵, 미나리, 숙주, 쇠고기를 한데 합하여 초간장을 넣고 살짝 버무린다.
- 접시에 담고 위에 김, 붉은 고추, 지단을 얹는다.

Ingredients

Main ingredients
Mung Bean Jelly 300g (4×0.5×0.5cm, julienne)

Sub-ingredients
Beef 50g (top round, julienne) · Korean Watercress 70g (stem only, cut into 4cm length)
Mung Bean Sprout 70g (trimmed, head & tail off)

Seasonings
For mung bean jelly · Sesame Oil 1/2t · Salt 1/4T
For beef · Soy Sauce 1/2T · Sesame Oil 1/2t · Sugar 1/2t · Garlic 1/2t (minced) · Green Onion (finely chopped) 1t · Sesame Seeds 1/2t · Black Pepper pinch
Vinegar soy sauce · Soy Sauce 1T · Sugar 1T · Vinegar 1T · Water 1T

Garnish
Egg, for Jidan 1ea (cut into 4×0.2cm) · Red Chili 1/2ea (julienne 2cm length) · Dried Seaweed 1/2 sheet (toasted and torn)

Steps

1. Preparing mung bean jelly
- Blanch mung bean jelly in boiling water and drain.
- Coat mung bean jelly with sesame oil.
- Add salt to taste.

2. Seasoning and stir-frying beef
- Marinate beef for 10 min.
- Stir-fry the beef.

3. Blanching vegetables
- Blanch Korean watercress and mung bean sprouts in boiling water and drain.
- Rinse with cold water and squeeze.
- Cut the Korean watercress into 4cm long pieces.
- Blanch mung bean sprouts in boiling water and drain.

4. Making vinegar soy sauce
- Mix all the ingredients to make vinegar soy sauce.

5. Tossing & plating
- Place mung bean jelly, Korean watercress, mung bean sprouts and beef in a bowl.
- Add vinegar soy sauce and toss gently.
- Transfer to plate and garnish with seaweed, egg jidan and red chili.

팥죽

붉은팥을 무르게 삶아 불린 쌀을 넣어 끓인 죽으로서 일 년 중 밤이 가장 긴 동지에 먹는
절기음식이다.

Pat-juk_ *Red Bean and Rice Porridge*

Pat-juk is a porridge made from red bean and soaked rice.
Koreans traditionally eat it on the winter solstice. the day when the night is the longest in the year.

재료

주재료
쌀 3/4C (2시간 이상 불리기) • 붉은팥 2C • 물 35C (=7리터)

부재료
찹쌀가루 1C • 더운물 2T (=찬물 3T)

양념
소금 1T • 설탕 1/2T

만드는 법

1. 팥 삶기
- 팥은 씻어서 냄비에 담고 물 5C을 부어 한소끔 끓인다.
- 물이 끓어오르면 바로 물을 따라 버리고, 다시 21C의 물을 붓고 끓인다.
- 팥물이 끓기 시작하면 중불로 줄인다.
- 물이 끓어오를 때마다 3C의 물을 3번 넣는다.
- 삶아낸 팥은 뜨거울 때 주걱으로 으깨고 체에 걸러 껍질은 버리고 앙금만 가라앉힌다.

2. 새알심 만들기
- 찹쌀가루는 뜨거운 물에 익반죽하여 지름 1cm 크기의 공을 빚는다.

3. 죽 끓이기
- 냄비에 팥 삶은 웃물과 불린 쌀을 넣고 저어가며 쌀알이 완전히 퍼질 때까지 끓인다.
- 가라앉힌 앙금을 넣고 저으면서 잘 섞이도록 끓이다가 새알심을 넣고 소금 1/2T를 넣고 간을 맞춘다.
- 새알심이 위로 떠오르면 소금으로 간한다.

4. 그릇에 담기 및 고명 올리기
- 죽그릇에 새알심을 고명 삼아 얹는다.
- 설탕과 소금을 곁들여 낸다.

Ingredients

Main ingredients
Rice 3/4C (Soaked minimum 2 hours) • Red Beans 2C • Water 35C (=7 liter)

Sub-ingredients
Sweet Rice Powder 1C • Hot Water 2T

Seasonings
Salt 1T • Sugar 1/2T

Steps

1. Preparing red beans
- Clean the red beans and boil it with 5cups of water for a few minutes.
- Drain the boiling water and add 21cups of water.
- When it gets boil, reduce the heat to medium.
- Add 3cups of water for 3 times.
- When the red beans get cook through, mash them with spatula. (Do this step with hot beans.)
- Strain the red bean water and red bean starch only. (remove the all skins.)
- Set a side. (Separate the red bean water and starch.)

2. Making saealsim (sweet rice ball)
- Mix sweet rice powder with hot water and knead them well.
- Make a saealsim as 1cm in diameter.

3. Boiling
- When red bean water gets rest, use the only top parts of red bean water.
- With that top water, boil soaked rice for 15 min.
- When it gets boil, add bottom parts of red bean water. (red bean starch)
- Stir well to prevent to burn.
- Add the saealsim with 1/2T of salt and boil it for 5 min.
- When saealsim gets rise in the pot, add 1/2T of salt.

4. Plating
- Transfer to a bowl and serve it with sugar and salt. (optional)

귀띔
Tips

1. 불의 세기에 따라 수분 증발량이 달라지므로 약불로 조절해 끓인다.
2. 죽에 어울리는 국간장, 소금, 맵지 않은 물김치(동치미나 백김치)와 함께 먹는다.
   한번에 많은 양을 담아내기보다는 조금씩 덜어서 먹을 수 있도록 앞 접시를 따로 곁들여 낸다.

1. Use low heat to control the amount of water in pat-juk.
2. Serve hot, with a small bowl to take some pat-juk out to eat.
   Juk goes well with gukganjang, salt and mul-kimchi. (dongchimi or baek-kimchi)

오이소박이

오이는 영양성분이 그리 많지 않지만 수분이 많고 상쾌한 향기와 씹히는 맛이 좋은 채소이다.
부추는 피를 맑게 해준다.

Oi-sobagi_ *Cucumber Kimchi Stuffed with Korean Chives*

With its refreshing, crunchy and juicy taste, oi-sobagi (cucumber Kimchi) is loved by Koreans, especially
in the summer. Since this Kimchi ferments easily, it needs to be refrigerated right after being made.
Well-fermented Oisobagi with cold noodles can be a great summer dish as well.

재료

주재료
오이 5개 (소금 1T, 물 1C)

부재료
부추 100g (잘게 썰기) · 양파 50g (잘게 썰기)

양념
고춧가루 3½T · 소금 1/2T · 설탕 1t · 마늘 (다진 것) 2t ·
파 (다진 것) 1T · 물 1T · 액젓 1T

만드는 법

1. 오이 손질
- 오이는 소금으로 문질러 깨끗이 씻는다.
- 오이는 3.5cm 길이로 자르고 가운데는 +로 칼집을 낸
 다.
- 소금물에 약 30분간 절인다.
- 절인 오이를 물에 헹궈서 물기를 짠다.

2. 부재료 손질하기
- 분량의 양념에 잘게 썬 양파를 넣고 잘 섞는다.
- 잘게 썬 부추를 넣고 섞는다.

3. 소 넣기
- 절여진 오이에 양념을 넣는다.

Ingredients

Main ingredients
Cucumber 5ea (Salt 1T, Water 1C)

Sub-ingredients
Buchu/Korean Chive 100g (chopped) · Onion 50g (chopped)

Seasonings
Red Chili Powder 3½T · Salt 1/2T · Sugar 1t · Garlic 2t
(minced) · Green Onion 1T (finely chopped) · Water 1T · Fish Sauce
1T

Steps

1. Preparing cucumber
- Rub cucumber with salt and wash.
- Cut cucumber into 3.5cm long pieces.
- Give cross-slits on one end, leaving another end intact.
- Soak them in brine for about 30 min. until softened.
- Rinse with water and squeeze lightly with hands.

2. Preparing stuffing
- Mix all seasonings in a bowl.
- Add Korean chive and onion and toss.

3. Stuffing
- Fill the stuffing on the slits of cucumber using hands and
 chopsticks.

239

귀띔
Tips
1. 절여진 오이는 물기 없이 꼭 짜야 아삭하다.
2. 부추에서 풋내가 날 수 있으므로 양념은 살살 버무린다.
3. 오이소박이는 빨리 익기 때문에 담근 후 바로 냉장고에 넣는다.
4. 차게 보관했다 먹으면 아삭하고 시원한 맛을 즐길 수 있다.
5. 국물이나 국수음식과 잘 어울린다.

1. Rub cucumber with salt to remove the tiny thorns on its skin.
2. Do not mix the stuffing too hard. Otherwise, it produces a grassy taste from the Korean chives.
3. Keep it immedietly in the refrigerator as oi-sobagi is fermented easily.
4. Serve cold to fully enjoy the crunchiness and refreshing taste.
5. It goes well with soups or nooddles.

무생채

무생채는 여러 가지 양념과 무가 잘 어우러진 시원한 반찬이다. 무의 본래 색을 강조하기 위해 종종 고춧가루 없이 만들기도 한다. 무의 시원한 맛은 고기의 맛과 대조되어 맛을 더욱 살려준다. 무에 들어 있는 소화효소는 단백질 분해를 돕는다.

Mu-saengchae_ *Fresh Radish Salad Tossed in Spicy Sour Dressing*

Mu-saengchae is a refreshing side dish which radish and various seasonings harmonize well with. It is often made without red chili powder to emphasize the original color of radish. The refreshing taste of radish not only contrasts with that of meat dishes but also enhances it. The digestive enzymes in radish help protein with disintegration.

재료

주재료
무 400g (채 썰기)

양념
고춧가루 2T · 설탕 2T · 식초 1½T · 소금 1/2T · 간장
1/4T · 다진 파 2t · 다진 마늘 1t · 참기름 1t

고명
통깨 2꼬집

만드는 법

1. 무 색들이기
- 채 썬 무를 고춧가루와 섞어 잠시 놓아둔다.
2. 양념 만들기
- 큰 그릇에 참기름을 제외한 양념을 모두 섞는다.
3. 버무리기
- 위에 준비한 재료들을 모두 잘 섞는다.
- 참기름을 넣고 살짝 버무린다.
4. 담아내기 및 고명 올리기
- 그릇에 옮겨 담고 통깨를 뿌려낸다.

Ingredients

Main ingredients
Radish 400g (julienne)

Seasonings
Red Chili Powder 2T · Sugar 2T · Vinegar 1½T · Salt 1/2T · Soy
Sauce 1/4T · Green Onion 2t (finely chopped) · Garlic 1t
(minced) · Sesame Oil 1t

Garnish
Sesame Seeds 2 pinches

Steps

1. Coloring radish
- Mix radish and red chili powder well and set aside for minutes.
2. Making dressing
- Mix all seasonings except for sesame oil in a big bowl.
3. Tossing
- Mix no. 1 and no. 2 well.
- Toss gently after adding sesame oil.
4. Plating and garnishing
- Transfer to a plate with pinches of sesame seeds on top.

귀띔
Tips

1. 무생채를 버무릴 때 너무 센 힘을 주면 무가 부스러진다.
2. 무를 양념과 버무리기 전에 색을 들여야 색이 곱다.
3. 무생채는 불고기를 비롯한 고기음식들과 잘 어울린다.

1. When tossing Mu-saengchae, strong hand-movement might make radish broken.
2. Coloring radish prior to seasoning aids to achieve beautiful color.
3. Mu-saengchae goes well with Bulgogi or many other meat dishes.

말미에

어린 시절, 집을 새로 지으며 구들장 놓는 것을 보았습니다.

큰 조각의 얇은 돌을 놓고 다시 조각조각 작은 돌을 맞춰가며 모양새를 만들어 얼추 되었다 싶으면 아궁이에 불을 지펴봅니다.

어느 한 곳에서 연기가 샌다 싶으면 이리저리 맞춰가며 돌려놓기를 수차례…

힘들어하는 인부들에게 술과 음식을 대접해 가며 지리한 작업을 이어가게 하곤 했지요.

몇 년간의 수업내용을 한 권의 책으로 내놓는 것이 이와 다르지 않아 잊었던 기억을 되새기게 합니다.

2008년 12월 26일. 중급 1과정 수료 후 삼청각에서 원장님을 모신 인연의 자리에서 늘 '한국음식을 어떻게 하면 세계인들에게 쉽게 이해시키고 접근하게 할 수 있을까?'를 고민하시던 원장님의 고견을 화두로 조금은 번거롭지만 우리 음식 원형의 조리법을 외국인에게 알리기 위한 토대 마련에 목표를 두고 시작한 것이 '한국전통음식 영어 입문반'이었습니다.

빨리 갈 길은 혼자 나서고 멀리 가는 길은 함께 가라는 옛말처럼 16명이 서로의 열정과 시간을 모아 조심스레 띄운 뗏목이 시간을 넘어오면서 여러 회원님들이 힘을 더하여 수업을 하였고 책으로 인사드리게 되었습니다.

잠시 여행 중에, 또는 길지 않은 머무는 시간을 쪼개어 그 나라의 음식과 문화를 배운다는 것이 쉽지만은 않았을 터인데도 수업 후 서툰 젓가락질로 음식을 먹으며 엄지를 세우고, 저마다 자기 나라의 음식이야기에 시간 가는 줄 모르고 서로의 연락처를 주고받으며 때론 젓갈에 따라 김치 맛이 다르냐는, 음양오행과 한국음식의 오방색에 관한 깊은 내용까지 질문해 가며 열의를 보여주던 외국인들의 모습에서 뿌듯함과 감사함, 그리고 우리를 다시 돌아보는 시간을 가질 수 있었습니다.

편편(片片)을 돌아보니, 궁중음식연구원에 누가 될세라 세세한 것까지 신경 쓰시며 허리수술 아랑 곳없이 젊은 사람들 수업 준비하게 하시고 화장실 구석구석 엎드려 닦으시던 지미재 회장님들의 모습은 눈물까지 돌게 했고 말 없는 교훈이 되었습니다. 개개인의 사정을 뒤로하고 토요일 영어반 수업에 열정을 쏟아주시던 여러분의 모습은 제가 '지미재' 회원임을 감사하게 해주었습니다.

소박한 바람이 있다면, 궁중음식연구원 지미재가 마치 밤하늘 한가운데서 사람들의 길잡이가 되어주는 것처럼 이 책이 세계 곳곳에서 수업하는 사람들에게 감동을 주었으면 좋겠습니다.

끝으로 시작은 같이하였으나 개개인의 사정상 마무리작업을 함께하지 못하신 김선복 · 이정숙 · 이향범 · 권혜진 · 김문예 · 이현숙 선생님께 아쉬운 마음을 전하며, 마음으로 함께한 대필의 글을 놓습니다.

243

2015년 3월
지미재 회원 이숙정

Afterthoughts and Acknowledgements

When I was little, I saw how people put in flat stones for floor heating system while building a Korean traditional house. First they put in large pieces of stones and then added smaller pieces to shape the structure. When the work was done, they started a fire in the furnace. If a wisp of smoke leaked out, they reshaped the structure to get it right, no matter how long it took. When the workers were getting exhausted, my mother offered them food and drinks to continue the hard work.

Creating this book reminds me of my childhood memory because it was not much different from that work; this book is based on our English cooking class that had been developed for four and half years. Putting all the works and experiences together into a book was not easy.

On December 26, 2008, our English cooking class, "Introductory Korean Food Cooking Class in English," was launched under the goal of teaching our original recipes to more foreigners. It was initiated by Bokryeo Han, President of the Institute of Korean Royal Cuisine, who always think about how the institute can help people around the world easily learn about and become familiar with Korean food.

There is a saying: "If you want to go fast, go alone. However, if you want to go far, go together." We wanted to go far; starting with 16 members, we worked together with passion and commitment. Thanks to our members' contribution, our English cooking class has been offered to many foreigners and finally this book has been created.

My sincere gratitude extends to all foreign learners who showed their strong interest in Korean food. Despite their brief stay in Korea, they took the time out of their busy schedule to learn about Korean food and culture. Although they were clumsy with chopsticks, they enjoyed Korean food after class, sharing their food culture. They even asked deep questions about Kimchi's different taste depending on its fermented ingredient; they were also intrigued by Yin—Yang theory and Five—Colors theory of Korean food. They made me reflect on our work and realize how it benefited all those foreign learners.

Also, I was deeply touched by the contribution and voluntary services of older Jimijae members. They gladly went extra mile to benefit our institute, paying attention to every single detail. They helped young members prepare for class, while cleaning a restroom despite their bad back. The young members also sacrificed their weekends and worked hard to provide a class every Saturday. All these selfless works made me proud that I am a member of "Jimijae."

My wish is that this book will promote Korean food to people around the world. Like the North Star, which stays the same in the sky and helps people find their way, I hope this book, which was made by Jimijae and the Institute of Korean Royal Cuisine, will also serve as an excellent guide for English—speaking people to Korean food and culture. I dream that the videos of people learning Korean food with this book will reach every corner of the world through YouTube and other SNS media.

Finally, my acknowledgements extends to the following members who started with us but couldn't be with us till the end: Sunbok Kim, Jungsook Lee, Hyangbum Lee, Haejin Kwon, Moonye Kim, and Hyunsook Lee.

I greatly appreciate all the hard work of people who helped in publishing this book.

March 2015

Sukjeong Lee, a member of Jimijae

245

찾아보기

Index

외국인도 빠져드는 한국밥상

247

Let's Plunge into Korean Cuisine

The first impression of the first edition published on 15 April 2015
The third impression of the first edition published on 10 August 2018

Publishing planner Jimijae, Institute of Korean Royal Cuisine
Written by Brenda Han/ Donghee Cho/ Eunkyung Lee/ Gyoungsoo Chen/ Jaeyoung Kim/ Joowon Chun/ Keumsoo Hur/ Kiboum Paik/ Kyeongsook Lee/ Kyungwha Han/Kyungyi Kim/Maesoon Kim/Mihyang Cheong/Mihyang Kim/ Misook Lee/Miyoung Lee/Myungeun Shin/Rockhun Kim/Seohyung Im/ Sooboo Lee/ Soojin Hur/ Soyoung Park/ Sujeong Seo/ Sukjeong Lee/ Taehyun Kim/ Wonil Lee/ Younghee Kim/ Youngok Shin/ Youngsoon Jeong
Publisher Wooksang Jin
Published by Baeksan Publishing Company
Photographer Donghyuk Choi
Food stylist Myungeun Shin, Seohyung Im
Model Younghee Kim
Proofreading Editorial Dept.
Design body text by Jungja Kang
Design cover by Jungeun Oh

Registration 1974. 1. 9. je 406-1974-000001 ho
Address Baeksan building 3th floor, 370, Hoedong-gil, Paju-si, Gyeonggi-do, Korea
Tel 02-914-1621
Fax 031-955-9911
E-mail edit@ibaeksan.kr
Hompage www.ibaeksan.kr

ISBN 979-11-5763-014-1 93590
Price 25,500won

저자와의
합의하에
인지첩부
생략

외국인도 빠져드는 **한국밥상**

2015년 4월 15일 초판 1쇄 발행
2022년 8월 30일 초판 4쇄 발행

기 획 궁중음식연구원 지미재
지은이 김경이, 김락훈, 김매순, 김미향, 김영희, 김재영, 김태현, 박소영, 백기범, 서수정,
　　　　신명은, 신영옥, 이경숙, 이미숙, 이미영, 이수부, 이숙정, 이원일, 이은경, 임서형,
　　　　전주원, 정미향, 정영순, 조동희, 천경수, 한경화, 한혜정, 허금수, 허수진
펴낸이 진욱상
펴낸곳 백산출판사
사 진 최동혁
푸드스타일리스트 신명은, 임서형
모 델 김영희
교 정 편집부
본문디자인 강정자
표지디자인 오정은

등 록 1974년 1월 9일 제406-1974-000001호
주 소 경기도 파주시 회동길 370(백산빌딩 3층)
전 화 02-914-1621(代)
팩 스 031-955-9911
이메일 edit@ibaeksan.kr
홈페이지 www.ibaeksan.kr

ISBN 979-11-5763-014-1 93590
값 25,500원